Plant Disease Management for Sustainable Agriculture

Sana Basharat
https://orcid.org/0000-0002-6737-5611
Hainan University, China

IGI Global
Scientific Publishing
Publishing Tomorrow's Research Today

Vice President of Editorial — Melissa Wagner
Director of Acquisitions — Mikaela Felty
Director of Book Development — Jocelynn Hessler
Production Manager — Mike Brehm
Cover Design — Jose Rosado

Published in the United States of America by
IGI Global Scientific Publishing
701 East Chocolate Avenue
Hershey, PA, 17033, USA
Tel: 717-533-8845 | Fax: 717-533-7115
Website: https://www.igi-global.com E-mail: cust@igi-global.com

Library of Congress Cataloging-in-Publication Data

Names: Basharat, Sana, 1992- editor
Title: Plant disease management for sustainable agriculture / edited by Sana Basharat.
Description: Hershey, PA : IGI Global Scientific Publishing, 2026. | Includes bibliographical references and index. | Summary: "This book offers a compilation of the latest research findings and case studies on plant disease management"-- Provided by publisher.
Identifiers: LCCN 2025034560 (print) | LCCN 2025034561 (ebook) | ISBN 9798337324296 hardcover | ISBN 9798337324302 paperback | ISBN 9798337324319 ebook
Subjects: LCSH: Artificial intelligence--Agricultural applications | Plant diseases | Agricultural innovations | Plants, Protection of
Classification: LCC S494.5.D3 P525 2026 (print) | LCC S494.5.D3 (ebook)
LC record available at https://lccn.loc.gov/2025034560
LC ebook record available at https://lccn.loc.gov/2025034561

British Cataloguing in Publication Data
A Cataloguing in Publication record for this book is available from the British Library.

Table of Contents

Preface xii

Chapter 1
Harnessing Genetic Engineering for Disease Resistance in Crops 1
Ashwini Talakayala, International Crops Research Institute for Semi-Arid Tropics, India
Raju Naik Sabavat, Sri Krishnadevaraya College of Horticultural Sciences, India
Kasi Rao Mediga, Godrej Agrovet Limited, India
Ravi Kiran Reddy Kondi, SPESS, LSU AgCenter, Louisiana State University, USA

Chapter 2
Genetic Engineering for Disease Resistance in Crops 35
Tapas Ranjan Das, ICAR- Indian Agricultural Research Institute, New Delhi, India
Neha Jha, Sardar Vallabhbhai Patel University of Agriculture and Technology, Meerut, India
Pooran Chand, Sardar Vallabhbhai Patel University of Agriculture and Technology, Meerut, India

Chapter 3
The Impact of Biodiversity on Plant Disease Dynamics in Organic Farming and Disease Management 71
Wasswa Shafik, Dig Connectivity Research Laboratory (DCRLab), Uganda

Chapter 4
Innovations in Disease Forecasting and Modelling 95
Deepak Gupta, Institute of Technology and Management, Gwalior, India
Apoorva Dwivedi, Galgotias University, India
C. Umarani, Christ Academy Institute for Advanced Studies, Bangalore, India
Rakhi Chawla, Asian Business School, Noida, India
S. M. Karpagavalli, Erode Sengunthar Engineering College, Thudupadhi, India
Someshwar Siddi, St. Martin's Engineering College, Dhulapally, India
Abduraimova Nigora, Termez University of Economics and Service, Termez, Uzbekistan

Chapter 5
Application of Remote Sensing in Early Disease Detection of Crop Plants 123
Ritesh Kalbande, Mohanlal Sukhadia University, Udaipur, India
Swati Hazra, Sharda University, Greater Noida, India
Manasvi Chand, Sharda University, Greater Noida, India

Chapter 6
Leveraging Deep Learning for Real-Time Medicinal Plant Identification 143
M. Martinaa, SASTRA University, India
G. Revathy, SASTRA University, India

Chapter 7
IoT-Based Fertilizer Pouring 169
Rajaa Vikhram Yohanandhan, SRM Institute of Science and Technology, India
P. Deepak, SRM Institute of Science and Technology, India
A. Ashik Ahamed, SRM Institute of Science and Technology, India
S. Vimal, SRM Institute of Science and Technology, India

Chapter 8
Web-Based Diagnostic Tool for Pulse Crop Disease Identification Using Convolutional Neural Networks 199
Ketan Bodhe, G.H. Raisoni College of Engineering, Nagpur, India
Kamlesh Ashokrao Kalbande, G.H. Raisoni College of Engineering, Nagpur, India
Dhananjay Bhagat, Department of Computer Science and Engineering (AI), Vishwakarma Institute of Technology, Pune, India
Bhagyashree Hambarde, St. Vincent Pallotti College of Engineering and Technology, Nagpur, India
Pranay Saraf, G.H. Raisoni College of Engineering, Nagpur, India
Asakti Rautkar, G.H. Raisoni College of Engineering, Nagpur, India
Ashvini Ganeshsingh Bais, G.H. Raisoni College of Engineering, Nagpur, India

Chapter 9
A Comprehensive Analysis Comparing Deep Learning's Performance in Plant Leaf Identification and Disease Detection 219
Shagufta Shakeel, Madan Mohan Malaviya University of Technology, Gorakhpur, India
Sushil Kumar Saroj, Madan Mohan Malaviya, University of Technology, Gorakhpur, India

Compilation of References 245

About the Contributors 275

Index 279

Detailed Table of Contents

Preface xii

Chapter 1
Harnessing Genetic Engineering for Disease Resistance in Crops 1
Ashwini Talakayala, International Crops Research Institute for Semi-Arid Tropics, India
Raju Naik Sabavat, Sri Krishnadevaraya College of Horticultural Sciences, India
Kasi Rao Mediga, Godrej Agrovet Limited, India
Ravi Kiran Reddy Kondi, SPESS, LSU AgCenter, Louisiana State University, USA

Plant diseases can critically undermine global food security due to the immense economic loss and yield reduction that they cause. Traditional breeding for disease resistance is time-consuming and limited by genetic diversity, highlighting the need for sustainable alternatives that reduce reliance on chemical pesticides. Precision genetic technologies provide sustainable alternatives by enabling the targeted incorporation of beneficial genes and genetic resources. This chapter summarizes recent advances in genetic engineering for developing disease-resistant crops, focusing on antimicrobial genes, RNA interference (RNAi), and CRISPR-Cas9–mediated genome editing for enhanced plant defense. In addition, the chapter discusses the effective use of genetically engineered disease resistance in major crops, along with potential risks and regulatory and policy considerations and highlights future research directions with sustainable approaches for durable disease management.

Chapter 2
Genetic Engineering for Disease Resistance in Crops 35
Tapas Ranjan Das, ICAR- Indian Agricultural Research Institute, New Delhi, India
Neha Jha, Sardar Vallabhbhai Patel University of Agriculture and Technology, Meerut, India
Pooran Chand, Sardar Vallabhbhai Patel University of Agriculture and Technology, Meerut, India

Genetic engineering has an advantage over traditional breeding to promote disease resistance in plants, which includes the ability to introduce, alter, or refine certain genes, swap genetic material between species, and re-engineer crops. GMOs are at the forefront of improving disease resistance in crops, providing sustainable means to agricultural adversity. Recent genome editing technologies like CRISPR/Cas9 have further transformed the sector, enabling accurate modifications like knocking out susceptibility genes or boosting resistance traits. Transgenic approaches put foreign genes, whereas genome editing provides the capacity to make precise changes to the plant's endogenous DNA without the addition of foreign DNA. The resistance obtained by RNAi is extremely long-lasting, and gene stacking produces resistance to several viral species. These innovations enhance crop productivity, reducing the use of chemical pesticides, and supporting sustainable agriculture. Genetic engineering can transform agriculture, ensuring sustainability and resilience against changing biological threats

Chapter 3
The Impact of Biodiversity on Plant Disease Dynamics in Organic Farming and Disease Management 71
Wasswa Shafik, Dig Connectivity Research Laboratory (DCRLab), Uganda

This chapter examines the role of biodiversity in shaping plant disease dynamics within organic farming systems and its implications for sustainable disease management. Biodiversity influences pathogen suppression, enhances soil microbial communities, and fosters ecological interactions that reduce disease incidence. It explores how diversified cropping systems, such as polycultures and crop rotations, contribute to disease resistance by disrupting pathogen life cycles and promoting beneficial microbial interactions. It also discusses the role of functional diversity in supporting natural enemies of plant pathogens, improving plant immunity through systemic resistance, and enhancing soil health. The chapter further evaluates how organic amendments, cover cropping, and habitat diversification affect plant-pathogen interactions and the effectiveness of biodiversity-driven disease management strategies. The discussion integrates current research, practical applications, and future perspectives on leveraging biodiversity to enhance resilience against plant diseases in organic farming.

Chapter 4
Innovations in Disease Forecasting and Modelling 95
Deepak Gupta, Institute of Technology and Management, Gwalior, India
Apoorva Dwivedi, Galgotias University, India
C. Umarani, Christ Academy Institute for Advanced Studies, Bangalore, India
Rakhi Chawla, Asian Business School, Noida, India
S. M. Karpagavalli, Erode Sengunthar Engineering College, Thudupadhi, India
Someshwar Siddi, St. Martin's Engineering College, Dhulapally, India
Abduraimova Nigora, Termez University of Economics and Service, Termez, Uzbekistan

Plant diseases pose significant threats to global food security, causing substantial crop losses worldwide. Traditional disease management approaches, primarily reactive in nature, are increasingly inadequate to meet the challenges of modern agriculture. The integration of machine learning (ML), deep learning (DL), and advanced predictive modeling technologies has revolutionized plant disease forecasting, offering proactive, accurate, and scalable solutions for disease prediction and management. This chapter provides a comprehensive review of innovative approaches to disease forecasting and modeling, examining the evolution from traditional epidemiological models to advanced machine learning-based predictive systems including temporal neural networks, ensemble forecasting methods, and hybrid climate-disease models. We analyze current forecasting methodologies, predictive datasets, validation frameworks, and real-world implementations while addressing key challenges such as uncertainty quantification, model interpretability, and long-term prediction accuracy.

Chapter 5
Application of Remote Sensing in Early Disease Detection of Crop Plants 123
Ritesh Kalbande, Mohanlal Sukhadia University, Udaipur, India
Swati Hazra, Sharda University, Greater Noida, India
Manasvi Chand, Sharda University, Greater Noida, India

Plant diseases cause significant damages to the crops and threatens global food security. Delay in detection and diagnosis of diseases often increase their severity and the extent of yield loss. Therefore, early disease detection is the need of the hour to save the crop production. Conventional methods of disease detection mostly relied on manual crop scouting and expertise of the farmers and researchers for the detection of the symptoms and diagnosis of the diseases as early as possible. However, there is often delay as the visible symptoms appear only after a certain stage of infection. Recent advances in hyperspectral remote sensing, unmanned aerial vehicles (UAVs), and artificial intelligence (AI) have created new opportunities for early disease detection and precise crop management. The current chapter critically explores the applications of remote sensing along with the deployment and evolution of modern AI/ML technologies in early plant disease detection, and future trajectories.

Chapter 6
Leveraging Deep Learning for Real-Time Medicinal Plant Identification 143
M. Martinaa, SASTRA University, India
G. Revathy, SASTRA University, India

The culture of medicinal plant cultivation has been in existence for many decades among the people of India. Most of the health benefit associated with medicinal plants comes from the Indian Forest. Medicinal plants have, therefore, been the subject of interest to scientists and researchers because of their importance in the survival of man. The identification of medicine plants requires a lot of time and the procedure must be done by experts in the field. Vision-based system can be easily and accurately used to identify the herb plants by scientists as well as the users. This research employs a deep learning approach to design an artificial intelligence system that can be used to identify the herb plants. In this paper, we have selected six plants from Kaggle database that holds maximum benefits which are Indian beech, curry, tulsi, betel, mint, and neem.

Chapter 7
IoT-Based Fertilizer Pouring 169
Rajaa Vikhram Yohanandhan, SRM Institute of Science and Technology, India
P. Deepak, SRM Institute of Science and Technology, India
A. Ashik Ahamed, SRM Institute of Science and Technology, India
S. Vimal, SRM Institute of Science and Technology, India

This report presents the design and implementation of an Internet of Things (IoT) system for automated fertilizer pouring in agricultural applications. The system leverages sensors, actuators, and cloud connectivity to optimize fertilizer application, enhance crop yield, and minimize environmental impact. Precisely delivers the required amount of fertilizer based on pre-programmed schedules or real-time sensor data. Enables farmers to remotely monitor system status, adjust settings, and receive alerts via a mobile app or web interface. Collects and analyzes sensor data to provide valuable insights into crop health, optimize fertilization strategies, and improve overall farm efficiency .Its excellent battery life ensures efficient operation without extended downtimes for recharging, maintaining the effectiveness of the spraying process. By offering an efficient solution, this agricultural robot represents a significant advancement in integrating modern technologies into farming, addressing operational challenges, and potentially transforming traditional agricultural practices.

Chapter 8
Web-Based Diagnostic Tool for Pulse Crop Disease Identification Using Convolutional Neural Networks 199
Ketan Bodhe, G.H. Raisoni College of Engineering, Nagpur, India
Kamlesh Ashokrao Kalbande, G.H. Raisoni College of Engineering, Nagpur, India
Dhananjay Bhagat, Department of Computer Science and Engineering (AI), Vishwakarma Institute of Technology, Pune, India
Bhagyashree Hambarde, St. Vincent Pallotti College of Engineering and Technology, Nagpur, India
Pranay Saraf, G.H. Raisoni College of Engineering, Nagpur, India
Asakti Rautkar, G.H. Raisoni College of Engineering, Nagpur, India
Ashvini Ganeshsingh Bais, G.H. Raisoni College of Engineering, Nagpur, India

Plant diseases threaten food security, especially in pulse crops essential for sustainable agriculture. This study explores Convolutional Neural Networks (CNNs) for rapid detection of major diseases such as Ascochyta blight, Fusarium wilt, and phytoplasma infections, which cause severe losses in countries like Ethiopia, India, and Taiwan. Using advanced imaging, we propose a web-based diagnostic tool to empower farmers with timely, accurate disease identification, enabling effective management decisions. This approach offers not just technological advancement but also economic sustainability for smallholder farmers by improving crop health and yields through early intervention and precision agriculture practices.

Chapter 9
A Comprehensive Analysis Comparing Deep Learning's Performance in Plant Leaf Identification and Disease Detection 219
Shagufta Shakeel, Madan Mohan Malaviya University of Technology, Gorakhpur, India
Sushil Kumar Saroj, Madan Mohan Malaviya, University of Technology, Gorakhpur, India

Agriculture is a significant element of the economy of a country, and the food that makes the nation healthier. Plant diseases should be determined to avoid wastage of resources and ensure the sustainability of the agricultural sector. Bacterial or animal diseases may seriously affect the food production system and the security of the global food supply. The application of deep learning has gained popularity because it has been used to diagnose and classify plant diseases within a very short period of time, on the basis of photos that are used.The study employs a hybrid deep learning model to predict the disease of bottle gourd and pumpkin leaves by the integration of EfficientNet, InceptionNet, and Transfer Learning. Our proposed models performed better in terms of precision and strength over the existing methods, with an accuracy of 83% in predicting the illness of pumpkin leaf and 96% in predicting the illness of bottle gourd leaf on a real picture dataset with augmentation methods.

Compilation of References 245

About the Contributors 275

Index 279

Preface

INTRODUCTION

Plant diseases remain one of the most formidable constraints to global agricultural productivity, threatening food security, farmer livelihoods, and ecological balance. In an era marked by climate variability, intensifying cultivation, and evolving pathogen populations, sustainable plant disease management has become both a scientific necessity and a societal imperative. The traditional reliance on chemical control measures, though effective in the short term, poses environmental, economic, and health concerns. Consequently, the integration of biotechnology, biodiversity-based approaches, artificial intelligence, and precision agriculture offers promising pathways toward resilient and sustainable farming systems.

Plant Disease Management for Sustainable Agriculture presents a multidisciplinary perspective on contemporary strategies designed to address plant health challenges through innovation and ecological stewardship. The volume brings together advances in genetic engineering, biodiversity-driven disease suppression, predictive modeling, remote sensing, deep learning, Internet of Things (IoT) technologies, and web-based diagnostics. Collectively, these contributions emphasize proactive, data-driven, and environmentally conscious approaches that align with the broader goals of sustainable agriculture.

This book aims to serve researchers, academicians, policymakers, extension professionals, and graduate students seeking to understand the evolving landscape of plant disease management. By bridging molecular biology with digital agriculture and ecological principles, the chapters highlight how integrated strategies can reduce chemical dependency, enhance crop resilience, and strengthen global food systems.

CHAPTER OVERVIEW

Chapter 1: Harnessing Genetic Engineering for Disease Resistance in Crops

This chapter explores precision genetic technologies as transformative tools for developing disease-resistant crops. It reviews antimicrobial gene deployment, RNA interference (RNAi), and CRISPR-Cas9–mediated genome editing as targeted strategies to enhance plant defense mechanisms. The discussion includes applications in major crops, biosafety considerations, regulatory frameworks, and future directions aimed at achieving durable and sustainable resistance.

Chapter 2: Genetic Engineering for Disease Resistance in Crops

Expanding on molecular approaches, this chapter compares transgenic technologies with advanced genome editing methods such as CRISPR/Cas9. It highlights gene stacking, susceptibility gene knockout strategies, and RNAi-mediated resistance for long-lasting disease control. Emphasis is placed on improving productivity while reducing pesticide reliance, thereby supporting resilient agricultural systems in the face of emerging biological threats.

Chapter 3: The Impact of Biodiversity on Plant Disease Dynamics in Organic Farming and Disease Management

This chapter examines the ecological foundations of disease suppression within organic farming systems. It discusses how crop diversification, polycultures, crop rotation, and soil microbial diversity influence pathogen dynamics. The role of functional biodiversity, organic amendments, and habitat management in strengthening plant immunity and soil health is critically analyzed, presenting biodiversity as a cornerstone of sustainable disease management.

Chapter 4: Innovations in Disease Forecasting and Modelling

Addressing the shift from reactive to predictive agriculture, this chapter reviews advancements in epidemiological modeling integrated with machine learning (ML) and deep learning (DL). It covers ensemble forecasting, temporal neural networks, hybrid climate-disease models, and validation frameworks. Key challenges such as uncertainty quantification, model interpretability, and long-term forecasting reliability are also examined.

Chapter 5: Application of Remote Sensing in Early Disease Detection of Crop Plants

Focusing on early diagnosis, this chapter explores hyperspectral imaging, unmanned aerial vehicles (UAVs), and artificial intelligence for real-time disease detection. It compares conventional scouting with advanced sensor-based systems and evaluates the integration of AI/ML tools for precision crop monitoring, highlighting future trajectories in remote sensing applications.

Chapter 6: Leveraging Deep Learning for Real-Time Medicinal Plant Identification

This chapter presents a vision-based deep learning framework for identifying medicinal plants. Using image datasets and convolutional neural networks, it demonstrates how artificial intelligence can support accurate, real-time recognition of valuable medicinal species, reducing dependency on expert-based identification and promoting digital botanical tools.

Chapter 7: IoT-Based Fertilizer Pouring

Introducing automation in nutrient management, this chapter details the design and implementation of an IoT-enabled fertilizer delivery system. It discusses sensor integration, cloud connectivity, real-time monitoring, and data-driven optimization of fertilization strategies. The system enhances resource-use efficiency, reduces environmental impact, and contributes to smart farming practices.

Chapter 8: Web-Based Diagnostic Tool for Pulse Crop Disease Identification Using Convolutional Neural Networks

This chapter presents a web-based diagnostic platform employing convolutional neural networks (CNNs) to detect major pulse crop diseases. By leveraging advanced imaging and AI-driven classification, the tool empowers farmers with rapid and accurate disease identification, supporting timely intervention and economic sustainability in pulse production systems.

Chapter 9: A Comprehensive Analysis Comparing Deep Learning's Performance in Plant Leaf Identification and Disease Detection

The final chapter evaluates hybrid deep learning architectures for plant leaf identification and disease prediction. Integrating EfficientNet, InceptionNet, and transfer learning approaches, the study compares model performance across datasets, demonstrating improved accuracy and robustness. The findings reinforce the growing role of deep learning in precision agriculture and automated disease diagnostics.

Together, these chapters underscore a unifying theme: sustainable plant disease management requires integration, of genetics and ecology, data science and field practice, innovation and regulation. By combining biological resilience with technological intelligence, this volume envisions a future where agriculture is not only productive but also environmentally responsible and adaptive to global change.

Chapter 1

Harnessing Genetic Engineering for Disease Resistance in Crops

Ashwini Talakayala
http://orcid.org/0000-0003-4528-4871
International Crops Research Institute for Semi-Arid Tropics, India

Raju Naik Sabavat
http://orcid.org/0000-0002-8629-477X
Sri Krishnadevaraya College of Horticultural Sciences, India

Kasi Rao Mediga
http://orcid.org/0000-0002-6817-6171
Godrej Agrovet Limited, India

Ravi Kiran Reddy Kondi
http://orcid.org/0009-0003-1052-035X
SPESS, LSU AgCenter, Louisiana State University, USA

ABSTRACT

Plant diseases can critically undermine global food security due to the immense economic loss and yield reduction that they cause. Traditional breeding for disease resistance is time-consuming and limited by genetic diversity, highlighting the need for sustainable alternatives that reduce reliance on chemical pesticides. Precision genetic technologies provide sustainable alternatives by enabling the targeted incorporation of beneficial genes and genetic resources. This chapter summarizes recent advances in genetic engineering for developing disease-resistant crops, focusing on antimicrobial genes, RNA interference (RNAi), and CRISPR-Cas9–mediated genome editing for enhanced plant defense. In addition, the chapter discusses the

DOI: 10.4018/979-8-3373-2429-6.ch001

effective use of genetically engineered disease resistance in major crops, along with potential risks and regulatory and policy considerations and highlights future research directions with sustainable approaches for durable disease management.

1. INTRODUCTION

To achieve global food security, it is essential to address the mounting challenge posed by plant diseases, which currently cause up to 30% potential annual losses in harvests. Although traditional breeding methods have been instrumental, they often become too time-consuming, often needing several years or even centuries to produce varieties resistant to diseases, and are hampered by a narrow genetic diversity. Such methods need to be complemented to respond to new and changing pathogens. A shift towards sustainable agriculture is key to ensuring long-term food security and protecting the environment. Techniques such as integrated pest management (IPM) effectively combine biological, cultural, and mechanical methods to develop resilient farming systems that rely on fewer chemical-based controls. pesticides. Genetic engineering is a powerful tool for sustainable control of diseases. Advanced techniques such as CRISPR/Cas9 and transgenics enable the fast and precise integration of disease resistance genes into crops. This technological advancement markedly decreases or eradicates the duration necessary for developing resilient varieties, thereby enabling scientists to respond promptly to emerging pathogens. Such a process consists of sustainability principles through a reduction of a reliance on chemical-based interventions and a resultant better-defined solution compared to classical breeding, often producing unpredictable effects. Lastly, the use of genetic engineering in crop development is crucial in creating a resilient food system capable of sustaining cumulative stresses emanating from climate change and mounting disease threats.

1.1 Impact of Plant Diseases on Global Food Security:

Plant diseases significantly impact global food security by reducing the output of crops, affecting food availability, and increasing economic insecurity, particularly in developing regions. Plant diseases, resulting from pathogens such as fungi, bacteria, and nematodes, result in enormous losses every year in both the quality and quantity of food crops (Singh et al., 2020; Gai & Wang, 2024). The global spread of plant disease is exacerbated by drivers such as climate change, which alters the pathogen evolution process and host-pathogen interaction, and global trade liberalization, which disseminates pathogens across the globe at an accelerated pace (Ristaino et al., 2021; Singh et al., 2023; Henz, 2024). Climate change affects not only how often and how severe plant diseases are but also shifts the distribution of pathogens, causing new outbreaks in regions that were previously unaffected (Chaloner et al., 2020; Mahanta et al., 2024). Traditional control measures, like chemical pesticides, are limited due to their harmful nature on the environment as well as the development of pathogen resistance (Ezrari et al., 2024; Islam et al., 2023). Consequently, sustainable and new approaches such as biological control, gene modification, and utilization of omics technology to enhance disease resistance in crops are in focus more and more (Islam et al., 2023). Quickly detecting and precisely identifying pathogens is vital for effective disease control, with molecular detection techniques and remote sensing being key tools in this process (Ezrari et al., 2024). Moreover, interdisciplinary integration and the application of eco-evolutionary theory in research are essential to improve future forecasting and control of plant disease epidemics in the face of climate change (Singh et al., 2023; Mahanta et al., 2024). Whereas plant diseases are still a significant global food security threat, science and international cooperation are still key to the evolution of sustainable ways through which their impact can be restricted and agricultural yield maintained (Ristaino et al., 2021; Henz, 2024).

1.2 Limitations of Traditional Breeding Approaches

Conventional plant breeding methods, although successful in former days to maximize crop yields and increase food production, have several limitations that hamper their effectiveness in solving modern agricultural problems. A chief limitation is a long duration to produce new cultivars ranging widely between one to two decades because of the complex process involving crossing, selection, and testing incorporated therein (Ahmar et al., 2020). Such a slow process fails to keep up with the rapid global demand for foods occasioned by rapid rises in populations (Ahmar et al., 2020). In addition, classical breeding techniques tend to create a decrease in genetic diversity; e.g., in breeding wheat, crossing between high-performing cultivars has resulted in a reduction in genetic variability and a plateauing in yield increments (Bedő & Láng, 2015). Usage of phenotypic selection in classical breeding further restricts effectiveness in choosing multi-dimensional traits valued by genotype-environment interactions like poor soil adaptation and soluble aluminum tolerance (Beaver & Osorno, 2009). Additionally, conventional breeding techniques have challenges crossing reproductive barriers and genetic drag capable of preventing introgression of valuable traits available in crop wild relatives (Pandey et al., 2019). Incorporation of genomic tools and novel biotechnology methodologies such as marker-assisted selection and genome editing presents tremendous possibilities in bridging these limitations via fast and accurate production of better crop varieties (Prohens et al., 2011; Pandey et al., 2019). However, utilization of these new strategies is usually hindered owing to a shortage of large-scale phenotyping facilities and a need to invest large sums in infrastructures and information (Bradshaw, 2017). Overall, although classical crop breeding has served as a basis for current agriculture, limitations inherent within it demand uniting novel methods to accelerate crop improvement and provide food security against fluctuating environmental and socio-economic conditions (Snape, 2004).

1.3 Importance of Sustainable Agriculture and Reduced Chemical Pesticides

Sustainable agriculture is gaining recognition as a crucial approach to address environmental and health issues associated with traditional farming methods, especially the extensive use of chemical pesticides. Reliance on synthetic pesticides has led to extensive environmental harm, including water contamination, soil pollution, and biodiversity loss, as well as negative health effects through the food chain (Seleiman et al., 2021; Peeriga et al., 2022). Sustainable agriculture endeavors to seek means of curtailing environmental impacts by adopting the preservation of natural resources and the promotion of healthy ecosystems. Such includes organic production systems, crop rotation, bio-fertilizers, and biochar, which promote soil fertility and reduce chemical input consumption (Seleiman et al., 2021). Biopesticides, which are natural substances like microorganisms and plants, provide an eco-friendly alternative to chemical pesticides. They are useful in integrated pest management (IPM) systems where they target pests specifically while leaving the beneficial organisms intact and ensuring that there is less chance of pest resistance (Peeriga et al., 2022; Ashiq, 2024). The adoption of biopesticides aligns with green chemistry standards and supports various Sustainable Development Goals (SDGs) by promoting ecological stewardship and food security (Fenibo et al., 2022). It improves resilience and biodiversity while reducing the frequency and severity of pest outbreaks through sustainable pest management practices such as the use of natural enemies and organic farming methods (Dey et al., 2024). International collaborations and policy initiatives play a vital role in advancing sustainable agriculture, which requires capacity building among farmers and stakeholders and effective knowledge sharing (Mali et al., 2023). Overall, sustainable agriculture addresses environmental and public health issues, economic growth, and the long-term resilience of the food system (Huzenko & Kononenko, 2024).

1.4 Role of Genetic Engineering in Sustainable Disease Management

Genetic engineering is at the core of sustainable disease management because it opens up new opportunities to create crop resistance and reduce the application of chemical pesticides, thereby promoting environmental sustainability. Genetic engineering technologies such as CRISPR-Cas9 have revolutionized the creation of disease-resistant crops by enabling precise and specific genetic modification. This has enabled the making of transgene-free, non-transgenic plants with resistance to various plant pathogens, which ultimately minimizes yield loss and pesticide use (Vincelli, 2016; Ali et al., 2022). Genetic engineering not only addresses the limitations of traditional breeding methods, which at times are time-consuming and labor-intensive, but also provides a dynamic solution to combat the constantly evolving threat of plant diseases (Uddin et al., 2024; Acharya et al., 2025). The integration of genetic engineering with other biotechnology tools, such as RNA interference and marker-assisted selection, adds to the potential to develop crops with broad-spectrum disease resistance (Islam et al., 2023). Furthermore, genetic engineering facilitates sustainable agriculture by improving the efficiency of nutrient use and developing stress-tolerant crops with the capacity to withstand environmental stresses such as drought and salinity (Lallawmkimi et al., 2024). Using genetic tools like gene drives and RNA interference for pest management provides sustainable options compared to chemical pesticides, reducing ecological impact and promoting biodiversity (Yadav et al., 2024). Despite the promising future of genetic engineering, there are still concerns about public perception, regulatory frameworks, and ethics that need open dialogue and sound risk assessments for safe and effective use (Lallawmkimi et al., 2024; Yadav et al., 2024). Finally, the combination of genetic engineering with traditional and modern agricultural practices holds immense possibilities for enhancing the resilience of crops, food security, and long-term sustainability in disease management (Ali et al., 2024; Ofori et al., 2025).

2. STRATEGIES FOR DEVELOPING DISEASE-RESISTANT CROPS THROUGH GENETIC ENGINEERING

Plant diseases caused by fungal, bacterial, viral, and insect pests pose a major threat to agricultural productivity and global food security (Gai and Wang, 2024). Most of the traditional breeding for disease resistance is often lowered by genetic variation and compatibility constraints. In this context, genetic engineering enables precise modification of plant genomes or insertion of novel resistance traits (Limenie and Alehegn, 2025), accelerating the development of disease-resistant crops and reducing chemical application. Modern biotechnological approaches such as transgenic development, RNA interference (RNAi), and genome editing technologies like CRISPR Cas9 for crop improvement, which majorly target the enhancement of plant defense mechanisms and open new sustainable and resilient agriculture. Some recent reviews highlight continuous progress in implementing engineering crop immunity and integrating advanced genomic strategies to address evolving pathogens in agriculture (Ali et al., 2022). Genes encoding antimicrobial proteins were introduced (antimicrobial peptides from other organisms) into crop genomes, giving plants a direct biochemical defense against bacterial, fungal, or viral pathogens. These proteins usually disrupt pathogen growth or survival, and a transgenic plant was developed to enhance resistance across variant pathogen types (Singh et al., 2025). RNA Interference (RNAi) uses a double-stranded RNA construct to trigger gene silencing in pathogens or pests by targeting their essential pathogen-related genes (e.g., viral replication proteins or pest survival genes). It gradually provides highly specific resistance, while reducing pesticide use. Instead, the application of host-delivered RNAi has been successful for virus and insect pest control in several crops (Saakre et al., 2024). Later, a new tool application, i.e., CRISPR technology, with its precise editing of plant immune genes or susceptibility factors (S-genes), enhances resistance without inserting foreign DNA; thus, genome editing can also alter or insert resistance (R) genes, creating broad-spectrum and durable disease. The high specificity and adaptability of CRISPR tools make them applicable for next-generation crop improvement (Saakre et al., 2024). In addition, all these genes and genetic resources are used to develop disease resistance traits from wild relatives or other species into elite crop varieties to produce. By stacking multiple resistance genes into one stack, which enhances durability against evolving pathogens. Implementation of modern genomics and bioinformatics helps to identify novel resistance genes, whereas genetic engineering technology has also been included in the expression of insecticidal proteins with neurotoxic effects (e.g., *Bt* Cry toxins derived from *Bacillus thuringiensis),* which are present within plants to protect against insect pests. These compounds act on insect guts and nervous systems but are engineered to be safe for non-targets like humans and insects, usually for widely grown *Bt* crops, which reduces pest damage and lowers pesticide requirements (Talakayala et al., 2020).

3. EFFECTIVE USE AND APPLICATIONS OF GENETICALLY ENGINEERED DISEASE RESISTANCE

Genetic engineering for disease resistance is used to enhance the natural defense mechanism in plants, genes encoding PRRs, and PR proteins with defense-related proteins, which are introduced into plants to enhance their innate immunity (Sharma et al., 2021). Genes that are involved in SAR (systemic acquired resistance) and RMA (RNA mediated silencing) can be introduced to enhance their acquired resistance (Kamle et al., 2020; Lopez-Gomollon & Baulcombe, 2022). This technology is used to introduce novel disease resistance genes into plants, where the mechanism of disease resistance in plants and how they can be enhanced through GE to develop new strategies for improving the health and productivity of crops are studied. It includes new emerging technology, *i.e.,* CRISPR/Cas9 technology investigated against viral, fungal, and bacterial diseases. Advancements in genetic engineering have enhanced the role of crop stress tolerance and overall agricultural productivity (Mackelprang et al., 2020). By improving various disease resistances in crops, i.e., banana plants were genetically engineered to express siRNA transgenic plants to develop resistance against viral replication initiation (Rep) genes, which showed resistance to Banana Bunchy Top Virus (BBTV) infection (Shekhawat et al., 2012). Usually, bacterial blight is caused by *Xanthomonas* oryzae in rice, where researchers targeted excision of the *OsSWEET13* and *OsSWEET14* promoters using CRISPR/Cas by developing resistant rice against blight (Zhou et al., 2015; Zeng et al., 2020). Recently, CRISPR/Cas9-mediated genome editing in the promoter region EBEs of the *OsSWEET14* gene has been demonstrated to confer resistance in Super Basmati rice lines against the Xoo strain carrying AvrXa7 TALE (Zafar et al., 2020). At the same time, the promoter region of *EBE-CsLOB* was edited using LbCas13a (Cpf1) and CRISPR/SpCas9P systems, resulting in enhanced resistance to citrus canker (Jia & Wang, 2020). In another study, Tripathi et al. (2021) developed banana lines resistant to *Xanthomonas wilt* (BXW) by editing the Musa dmr6 gene, a functional ortholog identified in Musa species. Overall, CRISPR-based genome editing has proven effective in conferring resistance against several devastating bacterial diseases in plants. Another study demonstrated that the expression of dual guide RNAs targeting the C1 gene (replication-associated protein) and the intergenic region (IR) of Cotton leaf curl Multan virus (CLCuMuV) using the CRISPR/Cas9 system conferred complete resistance to leaf curl disease (Yin et al., 2019). A multiplexed gRNA approach targeting multiple regions of the Chilli leaf curl virus (ChiLCV) genome conferred a high level of resistance in transgenic tobacco plants (Roy et al., 2019). Kis et al. (2019) demonstrated that an *Agrobacterium*-delivered plant transformation vector carrying four gRNAs targeting the movement protein (MP), coat protein (CP), long intergenic region (LIR), and N- and C-terminal regions of Rep/RepA of *Wheat dwarf virus (WDV)* conferred effective resistance, with transgenic plants remaining symptom-free. Tripathi et al. (2019) demonstrated that CRISPR-mediated inactivation of endogenous banana streak virus (*eBSV*; family *Caulimoviridae*) conferred resistance in banana, resulting in minimal symptoms compared with non-edited and transgenic control plants. Using this approach, CRISPR mutants did not confer resistance to African cassava mosaic virus in cassava (Mehta et al., 2019). Transgenic potato plants expressing Cas13a/sgRNA showed effective suppression of multiple strains of potato virus Y (PVY) with reduced disease symptoms (Zhan et al., 2019). Similarly, resistance to Southern rice black-streaked dwarf virus (SRBSDV) and Rice stripe mosaic virus was achieved in rice using the LshCas13a system (Zhang et al., 2019). The CRISPR/Cas9 system has been effectively used to confer resistance to multiple viral diseases, including tungro virus in rice (Macovei et al., 2018; Ramasamy et al., 2025). Fungal pathogens cause severe diseases in many economically important crops, and durable resistance is difficult to achieve due to the high polymorphism of pathogen effector proteins. In this context, CRISPR/Cas9-based approaches have emerged as promising tools for next-generation disease management in crop production. CRISPR/Cas9-mediated mutation of the native PMR4 gene in tomato reduced hyphal growth of *Oidium neolycopersici,* although complete resistance was not achieved (Martinez et al., 2020). Simultaneous CRISPR/Cas9 editing of *StDMR6-1* and *StDMR6-2* conferred late blight resistance in potato (Kieu et al., 2021). Enhanced resistance to *Phytophthora palmivora* was observed (Gumtow et al., 2018), and the pmr4 gene knockout increased resistance to powdery mildew in tomato. Knockout of *VvPR4b* activated defense responses against downy mildew caused by *Plasmopara viticola* (Li et al., 2020). Mutation of the Clpsk1 gene enhanced resistance to Fusarium oxysporum (FON), with loss-of-function mutants showing increased tolerance in watermelon seedlings (Zhang et al., 2020). Gao et al. (2021) reported that CRISPR/Cas9-mediated knockout of *SlYmiR482e-3p* in tomato enhanced resistance to *Fusarium oxysporum* f. sp. *lycopersici,* with increased expression of NBS–LRR resistance proteins. Therefore, genetic engineering has markedly improved crop disease resistance through the integration of transgenic approaches and CRISPR/Cas-based genome editing, enabling precise modification of resistance and susceptibility genes. These advances provide durable, broad-spectrum resistance, support the development of transgene-free crops, and accelerate breeding for effective disease management.

4. POTENTIAL THREATS, REGULATORY CONSIDERATIONS, AND POLICIES

Genetic engineering to develop disease resistant crops offers a powerful solution to enhance food security by reducing crop losses and reliance on chemical pesticides. However, this technology comes with prior potential threats and requires robust regulatory frameworks to ensure safety. Genetically modified (GM) crops pose various potential hazards, particularly regarding their long-term effects on the environment and ecosystems. One significant worry is gene flow, wherein engineered characteristics like resistance to herbicides or insects may transfer to wild species through cross-pollination. This occurrence, frequently referred to as "genetic contamination," could result in the emergence of "superweeds" that are hard to manage, ultimately jeopardizing the agricultural systems that this technology was designed to enhance (Thomas et al., 2022).

A major concern is the possibility of resistance failure. The extensive cultivation of crops featuring a single, universal resistance characteristic can impose significant selective pressure on pests and pathogens. As time progresses, these organisms may adapt to surpass the resistance mechanism, resulting in a new, more robust strain that makes the engineered trait ineffective (Perry et al., 2016). Beyond these direct effects, the widespread use of GE crops could also have broader biodiversity concerns. By promoting the use of a few, genetically uniform varieties, this technology may reduce the overall genetic diversity of crops. A lack of genetic variation makes the entire agricultural system more vulnerable to new diseases or environmental stresses. Finally, there are potential unintended ecological effects on non-target organisms. For example, some studies have raised concerns about the impact of insecticidal toxins expressed in GE crops on beneficial insects like bees and butterflies, highlighting the need for careful risk assessment and monitoring (Cellini et al., 2004; Kuzma et al., 2018). Secondly, threats related to health concerns, i.e., allergenicity and toxicity, and also antibiotic resistance. Allergenicity causes allergic reactions in susceptible individuals; most food allergies are associated with specific foods, and the introduction of new proteins through genetic modification could introduce new allergens. One example is inserting a gene from Brazil nuts into soybeans to enhance methionine content, which inadvertently introduced a known allergen, leading to allergic reactions in individuals sensitive to Brazil nuts. The project was halted with regulatory assessments, and testing of GM foods undergoes rigorous allergenicity tests, including in vitro assays and serum tests to ensure safety. Therefore, approval from GM food shows no increased allergy risk and is considered conventional food by the FDA and EFSA (EFSA, 2025). Antibiotic resistance is a concerning health issue, where GM crops developed using their marker genes facilitate the selection of successful modified crops/plants. Thus, practice has raised concerns about the potential transfer of these resistance genes to pathogenic bacteria; many GM crops now use non-antibiotic marker genes, and agencies continue to monitor ARMG (antibiotic resistance marker genes) risks (Gay et al., 2005). Lastly, it is concerned with socio-economic impacts, with market access and trade barriers among countries with stringent GM regulations that may impose trade restrictions, affecting the export of GM crops and also the involvement of IPR (intellectual property issues); patents on GM crops can lead to monopolies, potentially increasing seed costs for farmers and limiting access to technology (Van Esse et al., 2020) (Figure 1).

Figure 1. Potential Threats For Farmers

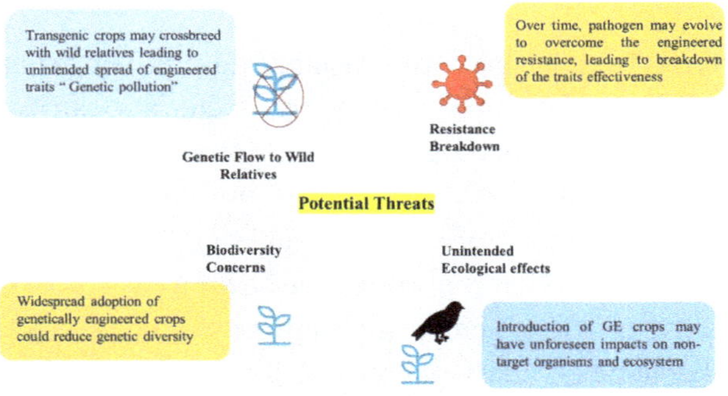

4.1 Regulatory Frameworks for Genetically Engineered Crops:

Genetically modified (GM) crops, created using contemporary biotechnology, offer considerable potential for boosting agricultural output, enhancing nutritional value, and offering resilience against biological and non-biological challenges. Nonetheless, their development and market introduction entail intricate scientific, environmental, and socio-economic factors. To validate that GM crops are safe for human health and the ecosystem, nations have put in place regulatory systems that oversee their evaluation, approval, and surveillance (Azeem et al., 2023). Regulatory systems for genetically engineered (GE) crops are designed to evaluate potential hazards, including ecological effects like gene flow and impacts on non-target species, as well as food safety issues such as allergenicity and toxicity. These regulations are influenced by international agreements, such as the Cartagena Protocol on Biosafety, and by national regulatory bodies, which may choose to implement either product-based or process-based methods based on the specific country. Moreover, regulatory frameworks are increasingly taking into account socio-economic aspects, public perception, and post-release management to successfully balance innovation with safety and public confidence (Nkhabindze et al., 2024).

As biotechnology continues to advance, particularly with the advent of gene-editing tools like CRISPR, regulatory frameworks are evolving to accommodate new techniques while maintaining rigorous safety standards. Understanding these regulatory systems is essential for developers, policymakers, and stakeholders to navigate the complexities of GE crop approval, commercialization, and sustainable deployment. In the USA, the Coordinated Framework for Biotechnology Regulation was established in 1986 and involves three federal agencies (Schiemann et al., 2019). Some of the regulatory considerations in the USA, Europe, and India are as follows: A) USA: 1. USDA (APHIS): Regulates the environmental release of genetically engineered organisms (Bratspies, 2012); FDA: Oversees the safety of food and feed derived from GE crops, and 3) EPA: Regulates the use of plant-incorporated protectants (PIPs), such as Bt proteins, for pest resistance (Grossman et al., 2019); therefore, its risk assessment is to evaluate potential environmental and health impacts before approving GE crops. B) European: 1) The precautionary principle in Europe requires extensive testing and approval processes for GE, with perception required for significant public resistance to GMOs, with regulatory policies and adoption rates, and C) INDIA: 1) Regulatory body: The Genetic Engineering Appraisal Committee (GEAC) under the Ministry of Environment, Forest, and Climate Change oversees the approval of GE crops such as Bt Cotton, only approved for commercial cultivation in India, for pest resistance. Therefore, several discussions continue regarding the approval of other GE crops, such as Bt brinjal, GM mustard (DMH-11 hybrid), and herbicide-tolerant crops, balancing innovation with environmental and health concerns (Choudhary et al., 2014; Padma et al., 2022; Shukla et al., 2018) (Table 1).

Table 1. Status of genetically engineered (GE) crops in India under the appraisal authority of the Genetic Engineering Appraisal Committee (GEAC), Ministry of Environment, Forest and Climate Change (MoEF&CC). The table summarizes regulatory decisions, commercialization status, and key references related to Bt cotton, Bt brinjal, and GM mustard (DMH-11).

Appraisal authority under MoEF&CC	Crop	Status in India	Reference
GEAC (Regulatory Body)	Bt Cotton	Commercialized in 2002; widespread adoption; strong safety and agronomic outcomes	Jayaraman, 2002; Qaim M, 2003
	Bt Brinjal	Approved in 2009; suspended in 2010; field trials approved 2020–2023	Herring RJ, 2014
	GM Mustard (DMH-11)	Tentative environmental release approved; pulled back pending further biosafety data	Yaduraju NT, 2021

As regulatory considerations, genetic engineering offers significant potential for enhancing disease resistance in crops; it is accompanied by various challenges and risks. Robust regulatory frameworks are essential to ensure that the benefits outweigh the potential threats. Ongoing research, transparent risk assessments, and public engagement are crucial for the responsible development and deployment of genetically engineered crops.

4.2. Policy Implications for Genetic Engineering in Agriculture

Genetic engineering in agriculture is a transformative advancement in modern crop science, which enables the development of crops with enhanced traits with innovations. These innovations hold the potential to address global challenges, including food insecurity, climate change, and sustainable agricultural production (Chaudhuri & Datta, 2018). The rapid advancement of GE also raises complex policy and regulatory queries with considerations of specific biosafety, environmental sustainability, socio-economic impacts, and ethical concerns. Policymakers usually balance the potential benefits of GE technologies with risks such as gene flow to wild relatives, unintended ecological consequences, public health considerations, and societal acceptance (Sukprasert et al., 2024).

As agricultural biotechnology advances, the development of effective policies and regulatory systems is vital for promoting responsible research, ensuring fair access, and safely incorporating GE crops into current farming practices. It is important for not just regulatory bodies but also scientists, farmers, and stakeholders to comprehend these policy implications as they contribute to shaping the future of sustainable agriculture (Hassan et al., 2024). Some of the policy implications of genetic engineering in agriculture are as follows.

1. **Risk Assessment and Biosafety Oversight**: Regulations should guarantee that genetically engineered (GE) crops designed for disease resistance undergo thorough assessments that consider environmental hazards (like gene transfer and effects on non-target organisms) as well as food safety issues such as potential allergenicity and toxicity. Different regulatory frameworks exist around the world—some nations adopt process-based systems that are activated by the techniques of genetic modification, while others implement product-based regulations that are triggered by the specific traits (Fernández et al., 2025).
2. **Intellectual Property & Farmers' Access:** The patent protection associated with GE seeds can limit access, especially for small-scale farmers, and influence fair distribution of benefits. It is crucial for policy frameworks to strike a balance between encouraging innovation and ensuring that products remain affordable and accessible, thereby securing widespread benefits for society.
3. **Socio-Economic Equity:** An effective policy should consider the expenses associated with regulatory compliance, particularly for smaller developers. Excessively strict regulations can hinder innovation and limit participation from public and small-sector researchers, consequently narrowing the variety of biochemical traits and crops that are developed (National Academies Press, 2016).

4. **Resistance Stewardship:** Traits for disease resistance can influence the adaptation of pathogens as time progresses. Regulations ought to require the implementation of resistance management strategies—like crop rotation, designated refuge zones, or stewardship practices—to maintain the efficacy of engineered resistances (Carrière et al., 2020).
5. **Public Perception & Biodiversity Protection**: Public apprehensions regarding biodiversity, cultural autonomy, and traditional seed practices shape regulatory choices. A pertinent example is the movement in Colombia against GMOs, which is propelled by indigenous communities aiming to safeguard native maize, underscoring the significance of public engagement and agricultural independence (Thomson et al., 2021).
6. **Evolving to Accommodate New Technologies:** New frontier technology like CRISPR requires new regulatory systems. Nations like India and China are implementing tiered, product-oriented strategies for specific genome edits, easing excessive regulations for precise, low-risk changes (Fernández et al., 2025).
7. **Research Investment & Public Sector Capacity:** To make sure that the advantages of genetically engineered (GE) crop innovations are accessible to all farmers, especially in low- and middle-income nations, it is crucial to have public investment and policies that promote fair technology transfer. This kind of support can help to spread the benefits more broadly, reaching beyond just the primary commercial crops and traits (Pixley et al., 2019).
8. **International Equity and Benefit-Sharing:** International agreements, such as the International Treaty on Plant Genetic Resources for Food and Agriculture, reinforce farmers' rights to maintain seed diversity and equitably participate in benefit-sharing, supporting ethical GE crop development (Esquinas-Alcázar et al., 2012) (Figure 2).

Figure 2. Schematic representation of policy implications of genetic engineering in agriculture.

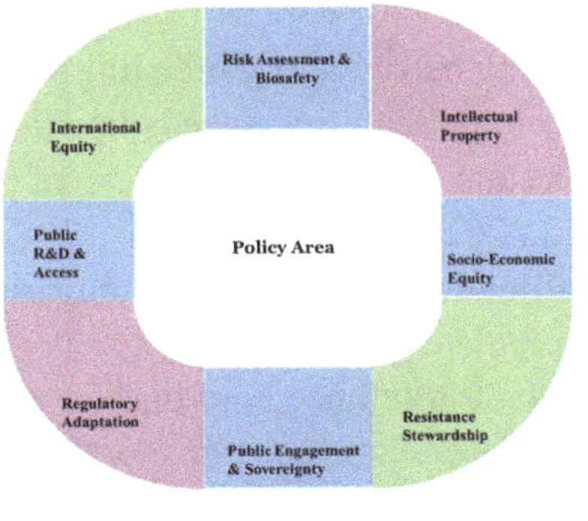

5. NEW FRONTIERS AND FUTURE DIRECTIONS

Integrated genetic engineering related to a sustainable approach combines with targeted molecular tools, *i.e.,* gene editing, transgenic strategies, and marker-assisted selection, with some of the agroecological practices; therefore, with improved agronomy, natural based solutions for an issue deliver resilient, productive, and low-input farming systems (Sharma et al., 2024; Collard & Mackill, 2008; Zaidi et al., 2019; Scheben et al., 2016; Altieri et al., 2015). Significant progress in genetic engineering, particularly through CRISPR/Cas-based genome editing (Zaidi et al., 2019; Gao, 2021), high-throughput genotyping (Varshney et al., 2021), and pangenome resources (Golicz et al., 2020; Tao et al., 2019), has created remarkable opportunities for developing crop varieties with specific, beneficial traits—such as better nutrient-use efficiency, resilience to drought and heat, resistance to diseases, and increased nutritional quality (Scheben et al., 2016; Qaim et al., 2020). Nevertheless, the potential of a genotype by itself does not ensure successful agricultural performance or sustainability. Actual improvements rely on how engineered genotypes interact with the agroecosystem, which includes soil health, microbial diversity, cropping practices, and management strategies (Bender et al., 2016; van Bruggen et al., 2019). Therefore, merging molecular precision with agroecological and management efforts an integrated genetic engineering method—presents the most promising route toward resilient, productive, and low-input farming systems (Bacha & Iqbal, 2023; Altieri et al., 2017; Wezel et al., 2020).

Integrated approaches combine targeted genetic modifications (such as alterations that decrease nutrient loss, enhance root foraging, or adjust hormonal responses) with supportive sustainable practices: conservation agriculture, cover cropping, crop diversification and rotations, enhancement of soil organic matter, precision management of nutrients and water, and inoculants based on microbiomes (Lal et al., 2020; Pretty et al., 2018; Glick et al., 2014). This collaboration guarantees that engineered traits are activated and utilized in varying field conditions, minimizes reliance on external chemical inputs, and boosts ecosystem services like carbon sequestration, soil structure, and pest control (Kassam et al., 2019; Wezel et al., 2020). Crucially, incorporating participatory breeding and involving farmers in the design process ensures local applicability, speeds up adoption, and tackles socio-economic challenges that frequently hinder the uptake of technology (Ceccarelli & Grando, 2007; Ojiewo et al., 2015).

Targeted Genetic Modification: Modern genetic engineering techniques have the ability to precisely manipulate genes controlling their agronomic traits, mainly nutrient use efficiency, drought tolerance, and disease resistance. 1) Nutrient use efficiency: Genetic edits target its nutrient transporter genes, *i.e.,* NRTs (nitrate) and PHTs (phosphates), enhance nutrient uptake and assimilation, reduce its fertilizer usage (Zaidi et al., 2019), and some of the root architecture modifications *via* genes like PHR1 improve soil exploration and phosphorus acquisition. 2) By using CRISPR mediated modifications of ABA signaling pathways, stomatal closure can be optimized, thus maintaining water-use efficiency under drought stress (Gao et al., 2021), heat stress by editing heat shock transcription factors (HSFs) and HSPs during high temperature by maintaining photosynthetic performance; 3) disease resistance-by targeting susceptibility genes (S-genes) and immune receptor genes (NLRs, PRRs) which target its genome for editing to enhance resistance against bacterial, fungal, and viral pathogens (Scheben et al., 2016) examples *OsSWEET* genes in rice reduces susceptibility to bacterial blight by preventing pathogen-mediated sugar efflux (Table 2).

Table 2. Focused genetic alterations to enhance important agronomic characteristics in crops. The table highlights key traits such as nutrient-use efficiency, drought and heat resilience, and disease resistance along with the associated target genes or pathways, the genetic engineering techniques employed (mainly CRISPR/Cas-based modifications), the results observed, and pertinent references.

S.No	Trait	Target Genes / Pathways	Genetic Engineering Approach	Outcome / Effect	References
1	Nutrient-Use Efficiency	NRTs (nitrate transporters), PHTs (phosphate transporters), PHR1	CRISPR/Cas-mediated gene edits, promoter modifications	Enhanced nutrient uptake and assimilation; reduced fertilizer dependence; improved phosphorus acquisition via modified root architecture	Zaidi et al., 2019
2	Drought Tolerance	ABA signaling pathway genes	CRISPR/Cas edits of ABA receptors and regulators	Optimized stomatal closure; improved water-use efficiency under drought stress	Gao, 2021
3	Root Architecture for Drought	DRO1, ARFs	CRISPR/Cas edits	Increased root depth and lateral branching; improved access to deeper soil moisture	Gao, 2021
4	Heat Tolerance	Heat shock transcription factors (HSFs), heat shock proteins (HSPs)	CRISPR/Cas-mediated edits	Maintained photosynthetic performance and growth under high temperatures	Gao, 2021
5	Disease Resistance	Susceptibility genes (S-genes), immune receptors (NLRs, PRRs), OsSWEET	CRISPR/Cas knockouts or promoter edits	Reduced susceptibility to bacterial, fungal, and viral pathogens; resistance without growth penalty	Scheben et al., 2017

5.1 Integration With Sustainable Practices:

Sustainable practices show that there is synergy between engineered genotypes and the use of sustainable agronomic practices; therefore, it is critical to maximize their potential by including conservation agriculture and soil health, agroecological diversification, precision nutrient and water management, and lastly, microbiome-based interventions (Table 3).

Table 3. Integration of sustainable agronomic practices with engineered genotypes to maximize crop performance. The table summarizes key practices, their approaches, synergistic effects with genetically improved crops, and references.

S.No	Practice / Focus Area	Description / Approach	Synergy with Engineered Genotypes / Expected Benefits	References
1	Conservation Agriculture & Soil Health	Reduced tillage, residue retention, crop rotations	Improves soil structure and microbial diversity; enhances nutrient cycling and stress resilience; supports full expression of engineered traits	Wezel et al., 2020
2	Agroecological Diversification	Intercropping, cover cropping, crop rotations	Buffers against pest outbreaks; improves ecosystem services such as pollination and carbon sequestration; complements stress-tolerant or nutrient-efficient genotypes	Wezel et al., 2020
3	Precision Nutrient & Water Management	Site-specific nutrient applications, optimized irrigation	Ensures nutrient-efficient varieties achieve maximum uptake and water-use efficiency; reduces input losses	—
4	Microbiome-Based Interventions	Application of beneficial microbial inoculants	Enhances nutrient acquisition, stress tolerance, and disease suppression; supports the performance of engineered genotypes	Glick, 2014

Under socio-economic considerations, farmers in breeding programs ensure that genetic innovation aligns with need (Ceccarelli & Grando 2007; Ojiewo et al., 2020). Future direction for maintaining sustainability. It includes a) multiplex trait stacking; b) integration with digital agriculture; c) scaling and policy support; and d) ecological monitoring and sustainability metrics.

a. **Multiplex Trait Stacking:** Editing several genes at the same time allows for the development of crop varieties that integrate drought resistance, disease resilience, and efficient nutrient use within one genotype. This approach of stacking traits speeds up the creation of robust, multi-stress-tolerant lines designed for agriculture that is smart about climate. Although thorough reviews are still being published, the applications of CRISPR are progressively facilitating simultaneous edits of multiple genes (Clapp & Ruder, 2020).
b. **Integration with Digital Agriculture:** The combination of gene-edited crops and digital farming innovations—such as sensors, remote sensing, predictive modeling, and farm-management systems—facilitates the tailored application of engineered traits in specific contexts. For instance, data from GIS, IoT, and drones is currently utilized to direct the introduction of CRISPR-enhanced crops into the most suitable agroecological regions. Additionally, the idea of the Internet of Everything (IoE) merges AI, IoT, molecular communication, and nanoscale sensing, allowing for precise monitoring and immediate decision-making at the levels of plants, soil, and molecules (Mamabolo et al., 2025; Babar & Akan, 2024).
c. **Scaling and Policy**: Developing integrated genetic engineering strategies necessitates strong collaboration between the public and private sectors, reformation of seed systems, and effective extension services to ensure that access is widespread and equitable. Public-private partnerships (PPPs) enable the combination of public research efforts with innovations from the private sector, promoting the dissemination of innovations and the commercialization of biotechnology traits in agriculture. For example, one of the studies from agricultural innovation networks, such as those in India, demonstrates how PPPs can establish digital infrastructures, data governance, and advisory systems that aid in the adoption of technology by smallholder farmers (Trotsenko & Slukin, 2020; Feuillet & Eversole, 2023; Harb et al., 2024).
d. **Ecological Monitoring and Sustainability Metrics:** Effective evaluation of long-term effects of the integration of genetic engineering through ecological observation where the research network is focused on long-term agroecosystems, *i.e.,* U.S. long-term agroecosystem research (LTAR), thus the network facilitates multi-decade investigations on soil, health, productivity, and ecosystems. Recently, one of the studies indicated that soil health indicators are strongly associated with yields in long-term trials with sustainable evaluations; another study on advanced remote sensing technologies such as WALGREEN forecasts soil organic carbon across space and time with monitoring soil health to ensure sustainable land management (Aroca-Fernandez et al., 2025) (Table 4).

Table 4. Future directions in integrating genetic engineering with sustainable agricultural approaches it highlights multiplex trait stacking, integration with digital agriculture, scaling through policy and partnerships, and ecological monitoring.

S.No	Future Direction	Description	References
1	Multiplex Trait Stacking	Edit multiple genes simultaneously for combined drought, disease, and nutrient-use efficiency traits.	Clapp et al., 2020).
2	Integration with Digital Agriculture	Combine gene editing with IoT, GIS, drones, and predictive modeling to optimize trait deployment in the field.	Mamabolo et al., 2025; Babar et al., 2024.
3	Scaling & Policy Support	Develop PPPs, seed system reforms, and extension models to scale integrated technologies equitably.	Trotsenko et al.,2020; Feuillet et al., 2023; Harb et al., 2024)
4	Ecological Monitoring & Metrics	Use long-term research networks, soil health metrics, and remote sensing tools to evaluate sustainability impacts of engineered systems.	Aroca-Fernandez et al., 2025)

5.2 Enhanced and More Enduring Compliance Models for Disease Management

For enhancing the successful disease management, it mostly relies not only on creating interventions but also on maintaining them over time. Conventional approaches, which are based on temporary enforcement or incentives, fail when certain pressures diminish. To create lasting effects, it is crucial to develop more sustainable and robust compliance frameworks. The following presents a structured exploration within key areas, backed by current evidence and recognized models:

1. Genetic and molecular compliance

   a. Genomic Editing for Resistance: By using genome editing for durable resistance, CRISPR/Cas knockout of susceptibility (S) genes, pyramiding of resistance (R) genes, and multiplex editing, plants can withstand rapidly evolving pathogens (Zaidi et al., 2019; Gao et al., 2021).
   b. Pathogen-Responsive Promoters: Synthetic promoters driving defense gene expression only during infection events to reduce penalties and prolong resistance.

2. Ecological and Agronomic Compliance: By implementing diversified cropping systems, rotating the crops' cultivations, and disrupting disease cycles and slow pathogen adaptation (Wezel et al., 2020), and later on by managing soil and microbiome health, which harnesses beneficial microbes (PGPRs, mycorrhizae,

biocontrol agents), improves plant immunity and suppresses soil-borne disease (Glick, 2014).

3. Technological and Digital Compliance: Early detection and forecasting, therefore using remote sensing and machine learning, and the implementation of portable diagnostics allow proactive management before epidemics. Precision agriculture platforms help farmers with some supportive tools, *i.e.,* fungicides, biocontrols, or using resistant cultivars in a timely and efficient manner. It also helps by using some of the early warning systems: digital DSS platforms such as Plantix and PlantVillage Nuru, *i.e.,* machine learning, to identify disease from images and provide real-time issues (Sama et al., 2023). By mapping through GIS based disease risk maps with crop improvement tools with localized interventions (fungicide spraying zones) (Mrisho et al., 2020; Kreuze et al., 2022).

For enhancing the genetic solutions, there will be some loss of effectiveness with pathogen evolution, seed/value chains spreading contaminated material, and social or regulatory barriers, as it requires an integrated compliance model that combines technical design (resistance stacking, monitoring) with institutional measures (stewardship, traceability, farmer engagement, incentives) (Van Esse et al., 2020; Ali et al., 2022). Some of the core elements of enhanced, enduring compliance/stewardship are mentioned in a tabular form (Table 5).

Table 5. Summary of key components of enhanced compliance and stewardship models for disease management in crop systems.

Component	Key Actions / Requirements	References
A. Pre-deployment	1. Multi-location efficacy trials & durability modeling 2. Risk assessment (gene flow, non-target effects, socio-economic impact) 3. Pre-release stewardship plan (developer + regulator)	Van Esse et al., 2020
B. Product Stewardship & Contracts	1. Clear user agreements (crop rotation, sanitation, reporting unusual disease) 2. Seed packaging with stewardship instructions & QR codes 3. Private-sector stewardship (like IRM/refuge programs)	Mbabazi et al., 2021
C. Traceability & Seed-Health Assurance	1.Digital tagging / lot IDs 2. Seed-health certificates (field/lab tests) 3. Public registry of approved varieties & stewardship rules	Blakeney et al., 2001

continued on following page

Table 5. Continued

Component	Key Actions / Requirements	References
D. Surveillance & Rapid Response	1. Sentinel surveillance (community scouts, remote sensing) 2. Routine pathogen monitoring (genotyping) 3. Pre-agreed mitigation protocols (roguing, seed replacement, stop-sowing windows)	Lynch et al., 2003
E. Farmer Engagement, Extension & Incentives	1. Participatory rollout (demo plots, plant clinics) 2.Incentives: subsidized certified seed, market access, crop insurance discounts 3.Training modules & local troubleshooting hotlines	World Bank Publications et al. 2023
F. Regulatory & Governance Fit	1.Transparent regulatory pathways for GE / gene-edited crops 2.Voluntary premarket consultation encouraged	Lemke et al., 2022
G. Resistance-Management Checklist	1.Stack complementary resistance mechanisms 2.Certified seed with stewardship label / lot ID 3.Synchronized planting/roguing 4.Monitor 3–5% of fields each season 5.Contingency plan for resistance break 6.Mandatory reporting of unusual outbreaks	Sisterson et al., 2013
H. Metrics to Measure Compliance	1.% seed lots with valid stewardship / traceability certificates 2.% farmers trained + adhering to stewardship 3.Time from outbreak report → action (days) 4.Frequency of virulence alleles in pathogens 5.Yield stability over 3–5 seasons	—

6. CONCLUSION

In conclusion, genetic engineering has transformed crop improvement by enabling precise, heritable modifications for targeted enhancement of agronomic traits. The integration of genetic engineering with sustainable agricultural practices represents a transformative pathway toward resilient, productive, and environmentally compatible farming systems. By combining molecular precision with ecological stewardship and participatory engagement, this approach addresses both biophysical and socio-economic challenges, offering a blueprint for future sustainable intensification of agriculture.

Publisher's Note

Correction Notice: A correction to this manuscript has been published

Reason: Published without tables

It was brought to our attention that this chapter was missing viable information pertaining to a few tables that were not present in the final version. We have since added the necessary tables for accuracy.

Correction Date: 27 March 2026

REFERENCES

Acharya, L. K., Manik, S., Pradhan, M., Birah, A., Khokhar, M. K., Majumder, D., & Hossain, A. (2025). Recent Advancements in Genetic Engineering Techniques in Disease Resistance of Crops: An Outlook. In *Plant Stress Tolerance* (pp. 99-112). CRC Press.

Ahmar, S., Gill, R. A., Jung, K. H., Faheem, A., Qasim, M. U., Mubeen, M., & Zhou, W. (2020). Conventional and molecular techniques from simple breeding to speed breeding in crop plants: Recent advances and future outlook. *International Journal of Molecular Sciences*, *21*(7), 2590. DOI: 10.3390/ijms21072590 PMID: 32276445

Ali, Q., Akhtar, T., Zuhra, N., Syarbiah, S., & Ega, A. A. (2024). Biotechnology for sustainable agriculture: Innovations in disease management. *Plant Prot*, *8*(4), 803–816. DOI: 10.33804/pp.008.04.5524

Ali, Q., Yu, C., Hussain, A., Ali, M., Ahmar, S., Sohail, M. A., Riaz, M., Ashraf, M. F., Abdalmegeed, D., Wang, X., Imran, M., Manghwar, H., & Zhou, L. (2022). Genome engineering technology for durable disease resistance: Recent progress and future outlooks for sustainable agriculture. *Frontiers in Plant Science*, *13*, 860281. DOI: 10.3389/fpls.2022.860281 PMID: 35371164

Altieri, M. A., Nicholls, C. I., Henao, A., & Lana, M. A. (2015). Agroecology and the design of climate change-resilient farming systems. *Agronomy for Sustainable Development*, *35*(3), 869–890. DOI: 10.1007/s13593-015-0285-2

Altieri, M. A., Nicholls, C. I., & Montalba, R. (2017). Technological approaches to sustainable agriculture at a crossroads: An agroecological perspective. *Sustainability (Basel)*, *9*(3), 349. DOI: 10.3390/su9030349

Aroca-Fernandez, J. M., Diez-Pastor, J. F., Latorre-Carmona, P., Elvira, V., Camps-Valls, G., Pascual, R., & Garcia-Osorio, C. (2025). A Collaborative Platform for Soil Organic Carbon Inference Based on Spatiotemporal Remote Sensing Data. *arXiv preprint arXiv:2504.13962*.

Ashiq, A. (2024). Biopesticides in Sustainable Agriculture: Enhancing Targeted Pest Control and Ecosystem Health. *Applied Agriculture Sciences*, *2*(1), 1–8.

Azeem, A., Ul-Allah, S., Sher, A., Ijaz, M., Sattar, A., Ahmad, W., & Qayyum, A. (2023). Biosafety and biosecurity in genetically modified crops. In *Sustainable agriculture in the era of the OMICs revolution* (pp. 501–510). Springer International Publishing. DOI: 10.1007/978-3-031-15568-0_23

Babar, A. Z., & Akan, O. B. (2024). Sustainable and precision agriculture with the internet of everything (IoE). *arXiv preprint arXiv:2404.06341*.

Bacha, S. A. S., & Iqbal, B. (2023). Advancing agro-ecological sustainability through emerging genetic approaches in crop improvement for plants. *Functional & Integrative Genomics*, *23*(2), 145. DOI: 10.1007/s10142-023-01074-4 PMID: 37133756

Beaver, J. S., & Osorno, J. M. (2009). Achievements and limitations of contemporary common bean breeding using conventional and molecular approaches. *Euphytica*, *168*(2), 145–175. DOI: 10.1007/s10681-009-9911-x

Bedő, Z., & Láng, L. (2015). Wheat breeding: current status and bottlenecks. In *Alien introgression in wheat: Cytogenetics, molecular biology, and genomics* (pp. 77–101). Springer International Publishing. DOI: 10.1007/978-3-319-23494-6_3

Bender, S. F., Wagg, C., & van der Heijden, M. G. (2016). An underground revolution: Biodiversity and soil ecological engineering for agricultural sustainability. *Trends in Ecology & Evolution*, *31*(6), 440–452. DOI: 10.1016/j.tree.2016.02.016 PMID: 26993667

Bradshaw, J. E. (2017). Review and analysis of limitations in ways to improve conventional potato breeding. *Potato Research*, *60*(2), 171–193. DOI: 10.1007/s11540-017-9346-z

Bratspies, R. M. (2012). Is Anyone Regulating: The Curious State of GMO Governance in the United States. *Vermont Law Review*, *37*, 923.

Carrière, Y., Brown, Z. S., Downes, S. J., Gujar, G., Epstein, G., Omoto, C., Storer, N. P., Mota-Sanchez, D., Søgaard Jørgensen, P., & Carroll, S. P. (2020). Governing evolution: A socioecological comparison of resistance management for insecticidal transgenic Bt crops among four countries. *Ambio*, *49*(1), 1–16. DOI: 10.1007/s13280-019-01167-0 PMID: 30903512

Ceccarelli, S., & Grando, S. (2007). Decentralized-participatory plant breeding: An example of demand driven research. *Euphytica, 155*(3), 349–360. DOI: 10.1007/s10681-006-9336-8

Cellini, F., Chesson, A., Colquhoun, I., Constable, A., Davies, H. V., Engel, K. H., Gatehouse, A. M. R., Kärenlampi, S., Kok, E. J., Leguay, J.-J., Lehesranta, S., Noteborn, H. P. J. M., Pedersen, J., & Smith, M. (2004). Unintended effects and their detection in genetically modified crops. *Food and Chemical Toxicology*, *42*(7), 1089–1125. DOI: 10.1016/j.fct.2004.02.003 PMID: 15123383

Chaloner, T. M., Gurr, S. J., & Bebber, D. P. (2020). The global burden of plant disease tracks crop yields under climate change. bioRxiv, 2020-04.

Chaudhuri, A., & Datta, A. (2018). Genetically modified (GM) crops: A potential source to combat global hunger and malnutrition. *Austin Journal of Nutrition and Food Sciences*, *6*(3), 1106.

Choudhary, B., Gheysen, G., Buysse, J., van der Meer, P., & Burssens, S. (2014). Regulatory options for genetically modified crops in India. *Plant Biotechnology Journal*, *12*(2), 135–146. DOI: 10.1111/pbi.12155 PMID: 24460889

Clapp, J., & Ruder, S. L. (2020). Precision technologies for agriculture: Digital farming, gene-edited crops, and the politics of sustainability. *Global Environmental Politics*, *20*(3), 49–69. DOI: 10.1162/glep_a_00566

Collard, B. C., & Mackill, D. J. (2008). Marker-assisted selection: An approach for precision plant breeding in the twenty-first century. *Philosophical Transactions of the Royal Society of London. Series B, Biological Sciences*, *363*(1491), 557–572. DOI: 10.1098/rstb.2007.2170 PMID: 17715053

Danso Ofori, A., Zheng, T., Titriku, J. K., Appiah, C., Xiang, X., Kandhro, A. G., Ahmed, M. I., & Zheng, A. (2025). The role of genetic resistance in rice disease management. *International Journal of Molecular Sciences*, *26*(3), 956. DOI: 10.3390/ijms26030956 PMID: 39940724

Dey, S., Saha, S., Mandal, I., & Nandi, R. (2024). Sustainable Pest Management: As an Alternative to Chemical Pesticides. *International Journal of Scientific Research in Engineering and Management*, *8*(10), 1–5. DOI: 10.55041/IJSREM38087

EFSA Panel on Genetically Modified Organisms (GMO), Casacuberta, J., Barro, F., Braeuning, A., de Maagd, R., Epstein, M. M., ... & Dumont, A. F. (2025). Current practice, challenges and future opportunities in the safety assessment of newly expressed proteins in genetically modified plants. *EFSA Journal*, *23*(8), e9568.

Esquinas-Alcázar, J., Hilmi, A., & Noriega, I. L. (2012). A brief history of the negotiations on the International Treaty on Plant Genetic Resources for Food and Agriculture. In *Crop Genetic Resources as a Global Commons* (pp. 135–149). Routledge.

Ezrari, S., Legrifi, I., Taoussi, M., Khadiri, M., Belabess, Z., & Lahlali, R. (2024). Plant–pathogen interactions and global food security. In *Plant pathogen interaction* (pp. 11–52). Springer Nature Singapore.

Fenibo, E. O., Ijoma, G. N., & Matambo, T. (2022). Biopesticides in sustainable agriculture: Current status and future prospects. *New and future development in biopesticide research: Biotechnological exploration*, 1-53.

Fernández Ríos, D., Quintana, S. A., Gómez Paniagua, P., Arrúa, A. A., Brozón, G. R., Bertoni Hicar, M. S., Castro Alegría, A., & Goberna, M. F. (2025). Regulatory challenges and global trade implications of genome editing in agriculture. *Frontiers in Bioengineering and Biotechnology*, *13*, 1609110. DOI: 10.3389/fbioe.2025.1609110 PMID: 40613005

Feuillet, C., & Eversole, K. (2023). An integrated, systems-wide approach is needed for public–private partnerships to drive genetic innovation in crops. *PLoS Biology*, *21*(7), e3002181. DOI: 10.1371/journal.pbio.3002181 PMID: 37410694

Gai, Y., & Wang, H. (2024). Plant disease: A growing threat to global food security. *Agronomy (Basel)*, *14*(8), 1615. DOI: 10.3390/agronomy14081615

Gao, C. (2021). Genome engineering for crop improvement and future agriculture. *Cell*, *184*(6), 1621–1635. DOI: 10.1016/j.cell.2021.01.005 PMID: 33581057

Gao, Y., Li, S. J., Zhang, S. W., Feng, T., Zhang, Z. Y., Luo, S. J., Mao, H.-Y., Borkovich, K. A., & Ouyang, S. Q. (2020). SlymiR482e-3p mediates tomato wilt disease by modulating ethylene response pathway. *Plant Biotechnology Journal*, *19*(1), 17–19. DOI: 10.1111/pbi.13439 PMID: 32621307

Gay, P. B., & Gillespie, S. H. (2005). Antibiotic resistance markers in genetically modified plants: A risk to human health? *The Lancet. Infectious Diseases*, *5*(10), 637–646. DOI: 10.1016/S1473-3099(05)70241-3 PMID: 16183518

Glick, B. R. (2014). Bacteria with ACC deaminase can promote plant growth and help to feed the world. *Microbiological Research*, *169*(1), 30–39. DOI: 10.1016/j.micres.2013.09.009 PMID: 24095256

Golicz, A. A., Bayer, P. E., Bhalla, P. L., Batley, J., & Edwards, D. (2020). Pangenomics comes of age: From bacteria to plant and animal applications. *Trends in Genetics*, *36*(2), 132–145. DOI: 10.1016/j.tig.2019.11.006 PMID: 31882191

Grossman, M. R. (2019). Genetic engineering in the United States: regulation of crops and their food products. In *Regulation of genome editing in plant biotechnology: A comparative analysis of regulatory frameworks of selected countries and the EU* (pp. 263–312). Springer International Publishing. DOI: 10.1007/978-3-030-17119-3_7

Gumtow, R., Wu, D., Uchida, J., & Tian, M. (2018). A Phytophthora palmivora extracellular cystatin-like protease inhibitor targets papain to contribute to virulence on papaya. *Molecular Plant-Microbe Interactions*, *31*(3), 363–373. DOI: 10.1094/MPMI-06-17-0131-FI PMID: 29068239

Harb, E. (2024, November). The Impact Of Public-Private Partnerships (Ppps) On Sustainable Agriculture And Food Security: A Comparative Analysis Of Different Models. In *ACE-International Conference* (pp. 298-301).

Hassan, B. (2024). The Impact of Agricultural Policy on Global Food Security. *Frontiers in Agriculture*, *1*(2), 325–355.

Henz, G. P. (2024). Impact of plant diseases on global food and nutritional security: A strategic theme for Brazil. *RAPP–Annual Review of Plant Pathology, Brasília, 30*, 7–28. DOI: 10.31976/0104-038321v300001

Herring, R. J. (2014). On risk and regulation: Bt crops in India. *GM Crops and Food: Biotechnology in Agriculture and the Food Chain*, *5*(3), 204–209. DOI: 10.4161/21645698.2014.950543 PMID: 25437239

Huzenko, M., & Kononenko, S. (2024). Sustainable agriculture: Impact on public health and sustainable development. *Health Economics and Management Review*, *5*(2), 125–150. DOI: 10.61093/hem.2024.2-08

. International Finance Corporation. (2023). *Working with smallholders: A handbook for firms building sustainable supply chains.* World Bank Publications.

. Islam, A., Elango, S., Saini, P. K., Jan, I., & Sharma, Y. (2023). Advancing plant pathology and the innovative methods for enhancing disease resistance. *Journal of Plant Biota*.

Jayaraman, K. S. (2002). India approves GM cotton. *Nature Biotechnology*, *20*(5), 415–415. DOI: 10.1038/nbt0502-415 PMID: 11981538

Jia, H., & Wang, N. (2020). Generation of homozygous canker-resistant citrus in the T0 generation using CRISPR-SpCas9p. *Plant Biotechnology Journal*, *18*(10), 1990–1992. DOI: 10.1111/pbi.13375 PMID: 32167662

Kamle, M., Borah, R., Bora, H., Jaiswal, A. K., Singh, R. K., & Kumar, P. (2020). Systemic acquired resistance (SAR) and induced systemic resistance (ISR): role and mechanism of action against phytopathogens. In *Fungal biotechnology and bioengineering* (pp. 457–470). Springer International Publishing. DOI: 10.1007/978-3-030-41870-0_20

. Kassam, A., Friedrich, T., Derpsch, R., & Kienzle, J. (2015). Overview of the worldwide spread of conservation agriculture. *Field Actions Science Reports. The Journal of Field Actions, 8.*

Kieu, N. P., Lenman, M., Wang, E. S., Petersen, B. L., & Andreasson, E. (2021). Mutations introduced in susceptibility genes through CRISPR/Cas9 genome editing confer increased late blight resistance in potatoes. *Scientific Reports*, *11*(1), 4487. DOI: 10.1038/s41598-021-83972-w PMID: 33627728

Kis, A., Hamar, É., Tholt, G., Bán, R., & Havelda, Z. (2019). Creating highly efficient resistance against wheat dwarf virus in barley by employing CRISPR/Cas9 system. *Plant Biotechnology Journal*, *17*(6), 1004–1006. DOI: 10.1111/pbi.13077 PMID: 30633425

. Kreuze, J., Adewopo, J., Selvaraj, M., Mwanzia, L., Kumar, P. L., Cuellar, W. J., ... & Blomme, G. (2022). Innovative digital technologies to monitor and control pest and disease threats in root, tuber, and banana (RT&B) cropping systems: Progress and prospects. *Root, Tuber and Banana Food System Innovations: Value Creation for Inclusive Outcomes*, 261-288.

Kumar, A., Mahanta, D., Dange, M. M., Trivedi, A., & Nandeha, N. (2024). Global challenges facing plant pathology: A review on multidisciplinary approaches to meet the food security. *Journal of Scientific Research and Reports*, *30*(6), 884–892. DOI: 10.9734/jsrr/2024/v30i62106

Kuzma, J. (2018). Regulating gene-edited crops. *Issues in Science and Technology*, *35*(1), 80–85.

Lal, R. (2020). Regenerative agriculture for food and climate. *Journal of Soil and Water Conservation*, *75*(5), 123A–124A. DOI: 10.2489/jswc.2020.0620A

Lallawmkimi, M. C., Veda, D. S., Yadav, A., Kumar, M., & Rout, A. (2024). Innovative approaches in crop genetic engineering for sustainable agriculture: A review. *Journal of Advances in Biology & Biotechnology*, *27*(8), 615–631. DOI: 10.9734/jabb/2024/v27i81177

Lemke, S. L. (2022). Gene Editing in Plants: A Nutrition Professional's Guide to the Science, Regulatory, and Social Considerations. *Nutrition Today*, *57*(2), 57–63. DOI: 10.1097/NT.0000000000000532

Li, M. Y., Jiao, Y. T., Wang, Y. T., Zhang, N., Wang, B. B., Liu, R. Q., Yin, X., Xu, Y., & Liu, G. T. (2020). CRISPR/Cas9-mediated VvPR4b editing decreases downy mildew resistance in grapevine (*Vitis vinifera* L.). *Horticulture Research*, *7*(1), 7. DOI: 10.1038/s41438-020-00371-4 PMID: 32922821

Limenie, A. D., & Alehegn, M. (2025). Genetic engineering for cereal crop yield improvement and disease resistant breeding. *TheScientificWorldJournal*, *2025*(1), 6743917. DOI: 10.1155/tswj/6743917 PMID: 40225354

Lopez-Gomollon, S., & Baulcombe, D. C. (2022). Roles of RNA silencing in viral and non-viral plant immunity and in the crosstalk between disease resistance systems. *Nature Reviews. Molecular Cell Biology*, *23*(10), 645–662. DOI: 10.1038/s41580-022-00496-5 PMID: 35710830

Lynch, R. E., Guo, B., Timper, P., & Wilson, J. P. (2003). United States Department of Agriculture–Agricultural Research Service research on improving host-plant resistance to pests. *Pest Management Science: Formerly. Pesticide Science*, *59*(6-7), 718–727. DOI: 10.1002/ps.636 PMID: 12846322

Mackelprang, R., & Lemaux, P. G. (2020). Genetic engineering and editing of plants: An analysis of new and persisting questions. *Annual Review of Plant Biology*, *71*(1), 659–687. DOI: 10.1146/annurev-arplant-081519-035916 PMID: 32023090

Macovei, A., Sevilla, N. R., Cantos, C., Jonson, G. B., Slamet-Loedin, I., Čermák, T., Voytas, D. F., Choi, I.-R., & Chadha-Mohanty, P. (2018). Novel alleles of rice eIF4G generated by CRISPR/Cas9-targeted mutagenesis confer resistance to Rice tungro spherical virus. *Plant Biotechnology Journal*, *16*(11), 1918–1927. DOI: 10.1111/pbi.12927 PMID: 29604159

. Mali, M., Latha, K. M., Sidar, S., & Kiran, S. (2023). Review of conservation agriculture practices for sustainable farming. *structure, 5*(6).

Mamabolo, E., Mashala, M. J., Mugari, E., Mogale, T. E., Mathebula, N., Mabitsela, K., & Ayisi, K. K. (2025). Application of precision agriculture technologies for crop protection and soil health. *Smart Agricultural Technology*, *12*, 101270. DOI: 10.1016/j.atech.2025.101270

Mbabazi, R., Koch, M., Maredia, K., & Guenthner, J. (2021). Crop biotechnology and product stewardship. *GM Crops and Food: Biotechnology in Agriculture and the Food Chain*, *12*(1), 106–114. DOI: 10.1080/21645698.2020.1822133 PMID: 33079624

Mehta, D., Stürchler, A., Anjanappa, R. B., Zaidi, S. S. E. A., Hirsch-Hoffmann, M., Gruissem, W., & Vanderschuren, H. (2019). Linking CRISPR-Cas9 interference in cassava to the evolution of editing-resistant geminiviruses. *Genome Biology*, *20*(1), 80. DOI: 10.1186/s13059-019-1678-3 PMID: 31018865

Mrisho, L. M., Mbilinyi, N. A., Ndalahwa, M., Ramcharan, A. M., Kehs, A. K., McCloskey, P. C., Murithi, H., Hughes, D. P., & Legg, J. P. (2020). Accuracy of a smartphone-based object detection model, PlantVillage Nuru, in identifying the foliar symptoms of the viral diseases of cassava–CMD and CBSD. *Frontiers in Plant Science*, *11*, 590889. DOI: 10.3389/fpls.2020.590889 PMID: 33391304

National Academies of Sciences, Medicine, Committee on Genetically Engineered Crops, Past Experience, & Future Prospects. (2016). Genetically engineered crops: experiences and prospects.

Nkhabindze, B. Z., Magagula, C. N., Earnshaw, D., Mhlanga, C. F., Matsebula, S. N., & Dladla, I. G. (2024). Regulatory framework for genetically modified organisms in the Kingdom of Eswatini. *GM Crops and Food: Biotechnology in Agriculture and the Food Chain*, *15*(1), 212–221. DOI: 10.1080/21645698.2024.2375664 PMID: 38963885

Ojiewo, C. O., Kugbei, S., Bishaw, Z., & Rubyogo, J. C. (2015). *Community Seed Production (Workshop Proceedings)* (pp. 1-176). ICRISAT.

Padma, T. V. (2022). When will India approve its first GM food crop? *Nature*, *611*(7937), 648–649. DOI: 10.1038/d41586-022-03738-w PMID: 36385294

Pandey, K., Dangi, R., Prajapati, U., Kumar, S., Maurya, N. K., Singh, A. V., & Rajan, R. (2019). Advance breeding and biotechnological approaches for crop improvement: A review. *International Journal of Chemical Studies*, *7*(1), 837–841.

Peeriga, R., Balagani, P. K., & Atmakuri, L. R. (2022). Prominence of Biopesticides to supersede chemical pesticides in sustainable agriculture. *International Journal of Pharmacy and Pharmaceutical Sciences*, *5*(1), 81–85.

Perry, E. D., Ciliberto, F., Hennessy, D. A., & Moschini, G. (2016). Genetically engineered crops and pesticide use in US maize and soybeans. *Science Advances*, *2*(8), e1600850. DOI: 10.1126/sciadv.1600850 PMID: 27652335

Pixley, K. V., Falck-Zepeda, J. B., Giller, K. E., Glenna, L. L., Gould, F., Mallory-Smith, C. A., Stelly, D. M., & Stewart, C. N.Jr. (2019). Genome editing, gene drives, and synthetic biology: Will they contribute to disease-resistant crops, and who will benefit? *Annual Review of Phytopathology*, *57*(1), 165–188. DOI: 10.1146/annurev-phyto-080417-045954 PMID: 31150590

Pretty, J., Benton, T. G., Bharucha, Z. P., Dicks, L. V., Flora, C. B., Godfray, H. C. J., Goulson, D., Hartley, S., Lampkin, N., Morris, C., Pierzynski, G., Prasad, P. V. V., Reganold, J., Rockström, J., Smith, P., Thorne, P., & Wratten, S. (2018). Global assessment of agricultural system redesign for sustainable intensification. *Nature Sustainability*, *1*(8), 441–446. DOI: 10.1038/s41893-018-0114-0

Prohens, J., Fita, A. M., Rodríguez-Burruezo, A., Raigón, M. D., Plazas, M., & Vilanova, S. (2011). Breeding for the Present and the Future: Achievements and Constraints of Conventional Plant Breeding and Contributions of Genomics to a New Green Revolution. *Bulletin of University of Agricultural Sciences and Veterinary Medicine Cluj-Napoca. Horticulture*, *68*(1).

Qaim, M. (2003). Bt cotton in India: Field trial results and economic projections. *World Development*, *31*(12), 2115–2127. DOI: 10.1016/j.worlddev.2003.04.005

Qaim, M. (2020). Role of new plant breeding technologies for food security and sustainable agricultural development. *Applied Economic Perspectives and Policy*, *42*(2), 129–150. DOI: 10.1002/aepp.13044

Ramasamy, S. P., Rathnasamy, S. A., Manickam, S., Chellappan, G., Natarajan, B., Swaminathan, M., Seshadri, G., & Muthurajan, R. (2025). Genome editing for improving disease resistance in rice (Oryza sativa L.)-Scope, achievements, limitations and future needs. *Physiological and Molecular Plant Pathology*, *138*, 102640. DOI: 10.1016/j.pmpp.2025.102640

Ristaino, J. B., Anderson, P. K., Bebber, D. P., Brauman, K. A., Cunniffe, N. J., Fedoroff, N. V., Finegold, C., Garrett, K. A., Gilligan, C. A., Jones, C. M., Martin, M. D., MacDonald, G. K., Neenan, P., Records, A., Schmale, D. G., Tateosian, L., & Wei, Q. (2021). The persistent threat of emerging plant disease pandemics to global food security. *Proceedings of the National Academy of Sciences of the United States of America*, *118*(23), e2022239118. DOI: 10.1073/pnas.2022239118 PMID: 34021073

Roy, A., Zhai, Y., Ortiz, J., Neff, M., Mandal, B., Mukherjee, S. K., & Pappu, H. R. (2019). Multiplexed editing of a begomovirus genome restricts escape mutant formation and disease development. *PLoS One*, *14*(10), e0223765. DOI: 10.1371/journal.pone.0223765 PMID: 31644604

Saakre, M., Jaiswal, S., Rathinam, M., Raman, K. V., Tilgam, J., Paul, K., Sreevathsa, R., & Pattanayak, D. (2024). Host-delivered RNA interference for durable pest resistance in plants: Advanced methods, challenges, and applications. *Molecular Biotechnology*, *66*(8), 1786–1805. DOI: 10.1007/s12033-023-00833-9 PMID: 37523020

Samal, I., Bhoi, T. K., Pradhan, A. K., & Mahanta, D. K. (2023). Plantix app: A success story of artificial intelligence in plant protection. *Van Sangyan, 24*.

Santillán Martínez, M. I., Bracuto, V., Koseoglou, E., Appiano, M., Jacobsen, E., Visser, R. G., Wolters, A.-M. A., & Bai, Y. (2020). CRISPR/Cas9-targeted mutagenesis of the tomato susceptibility gene PMR4 for resistance against powdery mildew. *BMC Plant Biology*, *20*(1), 284. DOI: 10.1186/s12870-020-02497-y PMID: 32560695

Scheben, A., Yuan, Y., & Edwards, D. (2016). Advances in genomics for adapting crops to climate change. *Current Plant Biology*, *6*, 2–10. DOI: 10.1016/j.cpb.2016.09.001

Schiemann, J., Dietz-Pfeilstetter, A., Hartung, F., Kohl, C., Romeis, J., & Sprink, T. (2019). Risk assessment and regulation of plants modified by modern biotechniques: Current status and future challenges. *Annual Review of Plant Biology*, *70*(1), 699–726. DOI: 10.1146/annurev-arplant-050718-100025 PMID: 30822113

Seleiman, M. F., & Hafez, E. M. (2021). Optimizing inputs management for sustainable agricultural development. In *Mitigating environmental stresses for agricultural sustainability in Egypt* (pp. 487–507). Springer International Publishing. DOI: 10.1007/978-3-030-64323-2_18

Sharma, A., Sharma, A., Kumar, R., Sharma, I., & Vats, A. K. (2021). PR proteins: key genes for engineering disease resistance in plants. In *Crop improvement* (pp. 81–98). CRC Press. DOI: 10.1201/9781003099079-7

Sharma, S. (2024). Organic agriculture for sustainable food systems: A comprehensive review of benefits and constraints. *Turkish Journal of Agriculture - Food Science and Technology*, *12*(8), 1476–1481. DOI: 10.24925/turjaf.v12i8.1476-1481.6482

Shekhawat, U. K., Ganapathi, T. R., & Hadapad, A. B. (2012). Transgenic banana plants expressing small interfering RNAs targeted against viral replication initiation gene display high-level resistance to banana bunchy top virus infection. *The Journal of General Virology*, *93*(8), 1804–1813. DOI: 10.1099/vir.0.041871-0 PMID: 22552945

Shukla, M., Al-Busaidi, K. T., Trivedi, M., & Tiwari, R. K. (2018). Status of research, regulations and challenges for genetically modified crops in India. *GM Crops and Food: Biotechnology in Agriculture and the Food Chain*, *9*(4), 173–188. DOI: 10.1080/21645698.2018.1529518 PMID: 30346874

Singh, B. K., Delgado-Baquerizo, M., Egidi, E., Guirado, E., Leach, J. E., Liu, H., & Trivedi, P. (2023). Climate change impacts on plant pathogens, food security and paths forward. *Nature Reviews. Microbiology*, *21*(10), 640–656. DOI: 10.1038/s41579-023-00900-7 PMID: 37131070

Singh, S., Sahoo, S., Mohapatra, R., Panigrahi, C. K., Daniel, G. R., Arunkumar, S., Sahoo, L., & Mohapatra, D. (2025). Advances in Genetic Engineering for Disease Resistant Crops: A Review. *Journal of Advances in Biology & Biotechnology*, *28*(6), 856–869. DOI: 10.9734/jabb/2025/v28i62447

Singh, V. K., Singh, R., Kumar, A., & Bhadouria, R. (2021). Current status of plant diseases and food security. In *Food security and plant disease management* (pp. 19–35). Woodhead Publishing. DOI: 10.1016/B978-0-12-821843-3.00019-2

Sisterson, M. S., & Stenger, D. C. (2013). Roguing with replacement in perennial crops: Conditions for successful disease management. *Phytopathology*, *103*(2), 117–128. DOI: 10.1094/PHYTO-05-12-0101-R PMID: 23075167

Snape, J. W. (2004, September). Challenges of integrating conventional breeding and biotechnology: a personal view. In *Proceedings of the 4th international crop science congress* (Vol. 26).

Sukprasert, K. R. I. T. C. H. A. I., & Phadungkit, W. I. C. H. A. I. (2024). Integrating sustainable development frameworks into agricultural policies: A policy analysis perspective. *Contemp. Issues Behav. Soc. Sci*, *11*(2), 1–12.

Talakayala, A., Katta, S., & Garladinne, M. (2020). Genetic engineering of crops for insect resistance: An overview. *Journal of Biosciences*, *45*(1), 114. DOI: 10.1007/s12038-020-00081-y PMID: 33051408

Tao, Y., Zhao, X., Mace, E., Henry, R., & Jordan, D. (2019). Exploring and exploiting pan-genomics for crop improvement. *Molecular Plant*, *12*(2), 156–169. DOI: 10.1016/j.molp.2018.12.016 PMID: 30594655

Thomas, R. (2022). Legal Challenges in Biosafety of Transgenic Plants. *Issue 6 Indian JL & Legal Rsch.*, *4*, 1.

Thomson, J. (2021). *GM crops and the global divide*. Csiro Publishing. DOI: 10.1071/9781486312665

Tripathi, J. N., Ntui, V. O., Ron, M., Muiruri, S. K., Britt, A., & Tripathi, L. (2019). CRISPR/Cas9 editing of endogenous banana streak virus in the B genome of Musa spp. overcomes a major challenge in banana breeding. *Communications Biology*, *2*(1), 46. DOI: 10.1038/s42003-019-0288-7 PMID: 30729184

Tripathi, J. N., Ntui, V. O., Shah, T., & Tripathi, L. (2021). CRISPR/Cas9-mediated editing of DMR6 orthologue in banana (Musa spp.) confers enhanced resistance to bacterial disease. *Plant Biotechnology Journal*, *19*(7), 1291–1293. DOI: 10.1111/pbi.13614 PMID: 33934462

Trotsenko, O., & Slukin, S. (2020). Public-Private Partnerships for agricultural innovation: opportunities and challenges in world practice. In *E3S Web of Conferences* (Vol. 176, p. 06004). EDP Sciences. DOI: 10.1051/e3sconf/202017606004

Uddin, M. M., Islam, N., & Halimuzzaman, M. (2024). Genetic engineering in crop improvement for diseases on resistance. *Int J Sci Healthcare Res*, *9*(3), 262–276. DOI: 10.52403/ijshr.20240332

van Bruggen, A. H., Goss, E. M., Havelaar, A., van Diepeningen, A. D., Finckh, M. R., & Morris, J. G.Jr. (2019). One Health-Cycling of diverse microbial communities as a connecting force for soil, plant, animal, human and ecosystem health. *The Science of the Total Environment*, *664*, 927–937. DOI: 10.1016/j.scitotenv.2019.02.091 PMID: 30769316

Van Esse, H. P., Reuber, T. L., & van der Does, D. (2020). Genetic modification to improve disease resistance in crops. *The New Phytologist*, *225*(1), 70–86. DOI: 10.1111/nph.15967 PMID: 31135961

Varshney, R. K., Bohra, A., Yu, J., Graner, A., Zhang, Q., & Sorrells, M. E. (2021). Designing future crops: Genomics-assisted breeding comes of age. *Trends in Plant Science*, *26*(6), 631–649. DOI: 10.1016/j.tplants.2021.03.010 PMID: 33893045

Vincelli, P. (2016). Genetic engineering and sustainable crop disease management: Opportunities for case-by-case decision-making. *Sustainability (Basel)*, *8*(5), 495. DOI: 10.3390/su8050495

Wezel, A., Herren, B. G., Kerr, R. B., Barrios, E., Gonçalves, A. L. R., & Sinclair, F. (2020). Agroecological principles and elements and their implications for transitioning to sustainable food systems. A review: Agroecological principles and elements and their implications for transitioning to sustainable food systems. A review. *Agronomy for Sustainable Development*, *40*(6), 40. DOI: 10.1007/s13593-020-00646-z

Yadav, S., Sarangi, S., Parmar, S., Thodusu, M., Soni, S., & Pandey, K. (2024). Genetic engineering in insect management: New frontiers in pest control. *Microbiology Research Journal International*, *34*(12), 106–121. DOI: 10.9734/mrji/2024/v34i121514

Yaduraju, N. T. (2021). The saga of genetically-modified (GM) and herbicide-tolerant (HT) crops in India. *Weeds-Journal of the Asian-Pacific Weed Science Society*, *3*(1), 38–55.

Yin, K., Han, T., Xie, K., Zhao, J., Song, J., & Liu, Y. (2019). Engineer complete resistance to Cotton Leaf Curl Multan virus by the CRISPR/Cas9 system in *Nicotiana benthamiana*. *Phytopathology Research*, *1*(1), 9. DOI: 10.1186/s42483-019-0017-7

Zafar, K., Khan, M. Z., Amin, I., Mukhtar, Z., Yasmin, S., Arif, M., Ejaz, K., & Mansoor, S. (2020). Precise CRISPR-Cas9 mediated genome editing in super basmati rice for resistance against bacterial blight by targeting the major susceptibility gene. *Frontiers in Plant Science*, *11*, 575. DOI: 10.3389/fpls.2020.00575 PMID: 32595655

Zaidi, S. S. E. A., Vanderschuren, H., Qaim, M., Mahfouz, M. M., Kohli, A., Mansoor, S., & Tester, M. (2019). New plant breeding technologies for food security. *Science*, *363*(6434), 1390–1391. DOI: 10.1126/science.aav6316 PMID: 30923209

Zeng, X., Luo, Y., Vu, N. T. Q., Shen, S., Xia, K., & Zhang, M. (2020). CRISPR/Cas9-mediated mutation of OsSWEET14 in rice cv. Zhonghua11 confers resistance to *Xanthomonas oryzae* pv. oryzae without yield penalty. *BMC Plant Biology*, *20*(1), 313. DOI: 10.1186/s12870-020-02524-y PMID: 32620100

Zhan, X., Zhang, F., Zhong, Z., Chen, R., Wang, Y., Chang, L., Bock, R., Nie, B., & Zhang, J. (2019). Generation of virus-resistant potato plants by RNA genome targeting. *Plant Biotechnology Journal*, *17*(9), 1814–1822. DOI: 10.1111/pbi.13102 PMID: 30803101

Zhang, D., Hussain, A., Manghwar, H., Xie, K., Xie, S., Zhao, S., Larkin, R. M., Qing, P., Jin, S., & Ding, F. (2020). Genome editing with the CRISPR-Cas system: An art, ethics and global regulatory perspective. *Plant Biotechnology Journal*, *18*(8), 1651–1669. DOI: 10.1111/pbi.13383 PMID: 32271968

Zhang, M., Liu, Q., Yang, X., Xu, J., Liu, G., Yao, X., Ren, R., Xu, J., & Lou, L. (2020). CRISPR/Cas9-mediated mutagenesis of Clpsk1 in watermelon to confer resistance to Fusarium oxysporum f. sp. niveum. *Plant Cell Reports*, *39*(5), 589–595. DOI: 10.1007/s00299-020-02516-0 PMID: 32152696

Zhou, J., Peng, Z., Long, J., Sosso, D., Liu, B. O., Eom, J. S., Huang, S., Liu, S., Vera Cruz, C., Frommer, W. B., White, F. F., & Yang, B. (2015). Gene targeting by the TAL effector PthXo2 reveals cryptic resistance gene for bacterial blight of rice. *The Plant Journal*, *82*(4), 632–643. DOI: 10.1111/tpj.12838 PMID: 25824104

Chapter 2
Genetic Engineering for Disease Resistance in Crops

Tapas Ranjan Das
http://orcid.org/0000-0001-6848-5161
ICAR- Indian Agricultural Research Institute, New Delhi, India

Neha Jha
Sardar Vallabhbhai Patel University of Agriculture and Technology, Meerut, India

Pooran Chand
Sardar Vallabhbhai Patel University of Agriculture and Technology, Meerut, India

ABSTRACT

Genetic engineering has an advantage over traditional breeding to promote disease resistance in plants, which includes the ability to introduce, alter, or refine certain genes, swap genetic material between species, and re-engineer crops. GMOs are at the forefront of improving disease resistance in crops, providing sustainable means to agricultural adversity. Recent genome editing technologies like CRISPR/Cas9 have further transformed the sector, enabling accurate modifications like knocking out susceptibility genes or boosting resistance traits. Transgenic approaches put foreign genes, whereas genome editing provides the capacity to make precise changes to the plant's endogenous DNA without the addition of foreign DNA. The resistance obtained by RNAi is extremely long-lasting, and gene stacking produces resistance to several viral species. These innovations enhance crop productivity, reducing the

DOI: 10.4018/979-8-3373-2429-6.ch002

use of chemical pesticides, and supporting sustainable agriculture. Genetic engineering can transform agriculture, ensuring sustainability and resilience against changing biological threats

1. INTRODUCTION

The world's growing population, at least 9.8 billion by the year 2050, presents unprecedented threats to crops. Increased food requirements can only be met by sharply higher production of crops, but this is undermined by disproportionate loss of yield due to microbial pathogens. Crop epidemics reduce crop yields by an estimated 15% due to bacterial and fungal plant pathogens, another 3% loss due to viruses. For these valuable crops like potatoes, microbial infection can lead to disastrous losses up to 30%. To combat the losses due to the diseases, genetic engineering has been a game-changer tool, providing accuracy and effective tools against plant diseases. Genetic engineering can enable scientists to introduce, modify, or silence certain genes that control disease resistance, even transferring desirable characteristics across species. Virus-resistant papaya in Hawaii and blight-resistant potatoes are some of the examples that can be achieved by genetically modified organisms (GMOs) for crop protection. Gene editing technologies, primarily CRISPR-Cas9, have gone a step further to revolutionize the disease resistant by site-specific plant DNA alteration without employing exogenous genes. These technologies decrease the production cost, hence making the agriculture industry more profitable and sustainable by minimizing chemical pesticides.

2. ECONOMIC AND YIELD LOSSES DUE TO PLANT DISEASES

Plants' diseases are of most significant concern to world agriculture production, due to huge economic and yield losses. The list of some devasting disease-causing pathogens and their impacts on plants is given below in Table 1.

Table 1. Different disease causing pathogens and their impacts on plants

Disease	Causal Organism	Pathogen Type	Impact on Plants	References
Powdery Mildew*(Wheat, Grapes, Cucurbits)*	*Erysiphe spp.*	Fungal	Manifests as white powdery fungal growth on foliage, restricting photosynthesis and reducing yield	(Sharma et al, 2024)
Late Blight*(Potato, Tomato)*	*Phytophthora infestans*	Oomycete	Causes rapid necrosis of leaves and decay of tubers, significantly affecting productivity	(Patel et al, 2022)
Leaf Rust*(Wheat, Barley)*	*Puccinia spp.*	Fungal	Produces orange-brown pustules on leaf surfaces, compromising photosynthetic efficiency and leading to yield loss	(Gupta et al, 2023)
Root Knot Disease*(Tomato, Potato, Brinjal, Okra, Melons)*	*Meloidogyne spp.*	Nematode	Forms characteristic root galls, obstructing nutrient absorption and stunting plant growth	(Reddy et al, 2024)
Mosaic Virus*(Tobacco, Tomato, Pepper)*	*Tobacco mosaic virus (TMV)*	Viral	Disrupts normal leaf pigmentation, causing mottled discoloration and reduced vigor	(Raj et al, 2025)

Devastating epidemics of plant disease have redraw agriculture, economies, and societies time and again throughout the centuries. The Irish Potato Famine (1845–1852), resulting from the oomycete *Phytophthora infestans,* destroyed as much as 90% of Ireland's potato crop – a food staple – and killed an estimated one million individuals and allowed huge emigration. Equally, the coffee rust epidemic (*Hemileia vastatrix*) wiped out Ceylon's coffee in the 1860s, reorganizing world beverage markets and consumption habits at its source (Chaloner et al, 2021). In the 20th century, Citrus Tristeza Virus (CTV) caused holocaustic losses, killing 75 million citrus trees in Brazil alone in the 1950s, with the current-day worldwide costs running over $1 billion a year (Agrios, 2005). Current viral pandemics threaten food security, for example, Banana Bunchy Top Virus (BBTV), causing stunting and fruit inhibition in bananas in 34 nations, threatening a crop that yields 400–600 kg/year per capita in Africa, and Maize Lethal Necrosis (MLN), a viral complex that resulted in 100% loss in Kenya's Rift Valley following its 2011 incursion through infected seed. Bacterial pathogens such as *Xylella fastidiosa* have destroyed historic olive groves in Italy's Puglia since 2013, killing 11 million trees and cutting olive oil production by 50%, causing economic harm worth more than €1 billion. Fungal diseases are no less damaging, of which wheat stem rust race Ug99, which appeared in Uganda in 1999, can potentially inflict 80% losses in a pan-continental wind-borne pathogen. Climate change magnifies these risks through changing pathogen distribution, virulence, and host susceptibility, and studies foreseen higher disease risks for 12 staple crops in a warmer world. Epidemics have the same causes: international trade spreads pathogens in contaminated planting stock (e.g., CTV on grafts, BBTV on suckers) agriculture monocultures heighten exposure, and control is undermined by scientific uncertainty, late containment, and inadequate resistant cultivars. Together, plant disease leads to more than $220 billion in annual crop losses that endanger food to increasing numbers of people.

3. CONVENTIONAL BREEDING VS GENETIC ENGINEERING

Traditional breeding and genetic modification are both strong crop and animal improvement tools, but with different limitations. Traditional breeding, performed on the principle of crossing and selection among organisms with favourable characteristics, is generally prolonged, several generations to make, occasionally decades. Also, it is subject to the gene pool available in sexually compatible species, a limitation that cannot permit inclusion of characters from incompatible species. The process may also be imprecise, in that unwanted genes may be passed along with wanted ones, so much backcrossing is necessary to eliminate unwanted characteristics. With its ancient history and widespread use, conventional breeding is still a slow, less reliable process than existing biotechnological methods (Figure 1).

Figure 1. The main key point difference between disease resistance by conventional breeding vs genetic engineering.

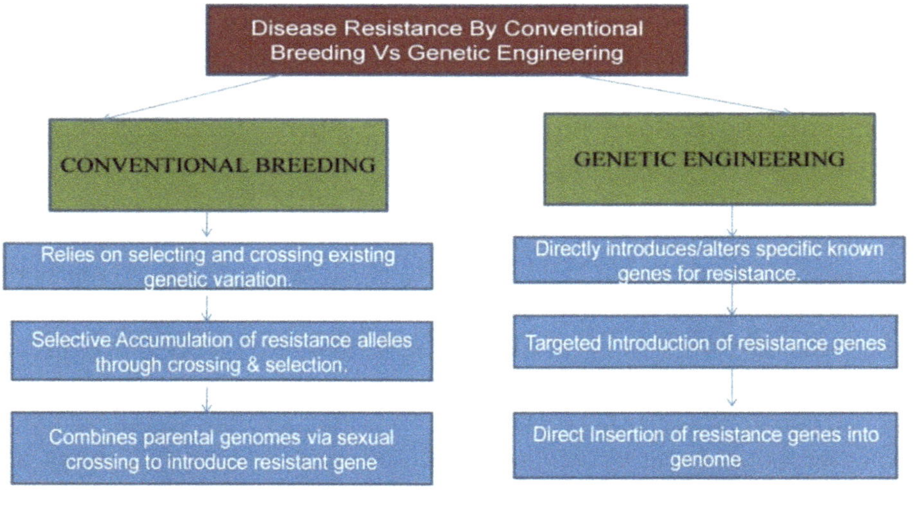

On the other hand, genetic modification offers the same specific alterations through direct addition, removal, or alteration of genes, making possible the introduction of traits from biologically disparate species. However, on its part, the procedure is not without problems, like the extremely great technical sophistication, requiring highly professional skills in molecular biology and bioinformatics. R and D and marketing clearance of genetically modified organisms (GMOs) are costly and time-consuming, including rigorous safety testing and public debate. Ethical concerns, potential environmental threats (e.g., gene spread among relatives in general), and consumer approval delay the adoption of genetically modified crops and animals. Even when genetic modification is quicker and more targeted than conventional breeding, it is still hampered by scientific, economic, and social limitations.

Genetic engineering has provided new opportunities for the production of disease-resistant crops and animals through specific genetic modification. One of them is transgenic technology, where a gene from another species is inserted to builtup resistance, such as pathogen-derived resistance, e.g., inserting viral genes (e.g., coat proteins) into plants to protect against infection, as has been realized in virus-resistant papaya. The second is introgression of natural resistance genes from wild relatives to susceptible crops to increases immunity. The CRISPR-Cas9 gene editing enables to selectively knock out susceptibility genes exploited by pathogens, for instance, mildew-resistant barley by *mlo* gene editing. Another technology, RNA interference (RNAi) is another option for pathogen gene knockdown, where by virus-resistant cassava and blight-resistant potatoes can be generated.

3.1 Limitations of Conventional Breeding in Developing Disease-Resistant Crop

Conventional plant breeding has been the mainstay of crop improvement for millennia. But in the case of the modern agricultural challenges especially the pressing demand for disease-resistant crops, its limitation has been felt increasingly intensely. We discuss hereinafter some of the key limitations:

3.1 1. Time Constraints

A fundamental limitation of conventional plant breeding lies in its inherently protracted timeframe. Developing improved varieties demands an extensive, multi-generational process. Breeders must meticulously perform cycles of crossing desirable parent plants, selecting offspring exhibit a fundamental limitation of conventional plant breeding lies in its inherently protracted timeframe. Developing improved varieties demands an extensive, multi-generational process and require 8 to 12 years for development of a new resistant variety.

3.1.2 Genetic Diversity Bottlenecks: Constraints on Disease Resistance Breeding

Most modern breeding programs are based on a high incidence of crossing among today's elite cultivars or highly related ones with no consideration for the vast reservoir of disease resistance genes existing these wild species having co-evolved together with pathogens under natural selection pressure in varied environments possess valuable resistance alleles that do not occur in the domesticated gene pool. Approximately 70% of the CWR species are in urgent need of collection and conservation, highlighting their underuse and vulnerability due to dependence on bred diversity. Gene erosion in domestic lines means in crop wild relatives (CWRs). breeders do not usually have the variation that they need to deal with strongly changing pathogens. Linkage Drag is another cause, although CWRs possess favorable resistance genes, their transmission into elite cultivars is the most acute linkage drag issue. This is an occurrence in the fact that the region of the chromosome bearing the desired resistance gene also bears many tightly linked alleles which regulate unwanted wild characteristics, i.e., low yields, poor-quality fruit, small seed, bitter flavours, or ill-timed flowering . The introduction of resistance from wild bananas (*Musa acuminata* ssp. *malaccensis*) to market varieties, for example, usually also introduces undesirable traits poorly suited to commercial culture or consumer taste. To get rid of these undesirable characteristics involves frequent backcrossing to domestic type over a number of generations—something stupidly time-consuming and capital-hungry in perennials such as bananas or fruit trees because of prolonged generation cycles and huge costs of holding plants to maturity for phenotypic assessment. It massively retards deployment of resistant varieties. Another important bottle neck is Monoculture Susceptibility i.e. union of narrow genetic foundation and the recalcitrance of incorporating new resistance genes has resulted in single, frequently dominant, resistance genes utilized in extensive monocultures. This applies immense selection pressure to pathogens to breach the resistance. A standout example is the Sr31 gene for wheat stem rust resistance. Its near unanimity in use provided adequate protection for decades, but the emergence of the Ug99 race of stem rust in Africa rendered it obsolete, threatening global wheat output. This pattern deplorably repeats throughout agricultural history. The 1840s Irish Potato Famine was caused by the virtual genetic homogeneity of Irish potatoes, all of which were vulnerable to *Phytophthora infestans* (NRC, 1993). Likewise, the initial Panama disease (Fusarium wilt TR1) epidemic devastated the world banana economy in the mid-20th century by killing off the susceptible 'Gros Michel' monoculture bananas (Ploetz, 2015). The crop has been replaced by resistant Cavendish clones, but history repeats itself with the identification and dissemination of Tropical Race 4 (TR4), which breaks Cavendish resistance and threatens the prevailing export banana. The

Cavendish monoculture, with a presence on the majority of export plantations, is the peak of the high genetic susceptibility generated by reliance on one clonally propagated variety with monomorphic resistance mechanisms. Genome studies have validated that such population bottlenecks decrease overall genetic diversity substantially and enhance noxious variation, further elevating susceptibility.

3.1.3. Climate Change Challenges

Climate change significantly worsens threats from plant diseases by disrupting pathogen behaviour and plant-pathogen relationships. Conventional breeding struggles to adapt effectively due to several key factors: Shifting Pathogen Ranges, where warmer conditions and altered precipitation patterns enable diseases to establish in previously unaffected areas (e.g., the spread of wheat blast from South America into parts of Asia); Combinatorial Stress Interactions, requiring plants to possess combined resistance to both diseases and simultaneous environmental pressures like drought or heat, a complex trait pyramid difficult to assemble economically using traditional methods; and Unpredictable Pathogen Evolution, where pathogens adapt more swiftly to changing climates than breeders can develop and deploy new resistant varieties, resulting in a persistent "resistance gap".

4. THE PROMISE OF GENETIC ENGINEERING

Genetic engineering presents possibilities of overcoming the limitations of conventional breeding through providing precise, rapid, and responsive solutions for generating disease-resistant plants. Recent technologies like transgenic, RNAi, gene stacking, gene editing, CRISPR-Cas9 which enable precision modifications, such as knocking out susceptibility genes (e.g., *mlo* in barley against powdery mildew) or boosting the function of resistance (R) genes, without the "linkage drag" (Table 2). CRISPR-edited rice with blast resistance was achieved in three years and transgenic papaya resistant to Papaya Ringspot Virus (PRSV) was developed in eight years. Editing wheat genes to counter emerging stem rust strains like Ug99, and enables synergy with climate adaptation by stacking disease resistance with traits like drought tolerance are some of the promising examples. Achievements such as powdery mildew-resistant wheat and blight-resistant potatoes highlight its capacity to limit chemical inputs and stabilize yield. With climate change posing enhanced disease pressure, the precision, speed, and flexibility of genetic engineering place it as a critical resource for the guarantee of global food security and agriculture system development that is climate resilient.

Table 2. Traditional breeding VS Genetic Engineering

Aspect	Traditional Breeding	Genetic Engineering	References
Methodology	Involves selective crossbreeding over multiple generations	Direct genetic alterations using advanced tools like CRISPR-Cas9	Sadhu & Chatterjee, 2025
Precision and Speed	Slow and largely unpredictable, requiring years for stable traits	Highly targeted and efficient, introducing desired traits in one generation	Jones & Dangl, 2006
Source of Resistance Genes	Restricted to natural genetic diversity within species or closely related relatives	Can incorporate resistance genes from diverse sources, even unrelated organisms	Trindade et al, 2021
Genetic Stability	Risk of undesired traits due to genetic linkage drag	Precision editing minimizes unwanted traits while preserving desirable qualities	Klindworth et al, 2023; Huang et al, 2023
Effectiveness Against Pathogens	Resistance can be overcome due to long breeding cycles	Continuous adaptability allows improved resistance against evolving pathogens	Mahanayak, 2024; Suhag, 2024
Ethical and Regulatory Considerations	Generally accepted with minimal regulatory restrictions	Faces scrutiny due to ethical concerns and requires biosafety evaluations	Kumar, 2024
Environmental Impact	Often relies on chemical treatments to sustain resistance	Helps reduce reliance on pesticides by integrating genetic resistance	Islam et al, 2024

5. MECHANISMS OF DISEASE RESISTANCE R-GENE BREEDING AND PR PROTEIN OVEREXPRESSION IN PLANTS

- Important mechanism for disease resistance are **Innate Immunity: Pump Triggered Immunity(PTI),** Effector-Triggered Immunity (ETI) and Systemic Acquired Resistance (SAR) Fig. 2
- Crops are engineered with stacked R-genes to combat evolving pathogens (e.g., wheat stem rust resistance).
- Transgenic plants with enhanced PR protein production show improved disease resistance (e.g., rice with chitinase genes).

Figure 2. Important mechanism for disease resistance

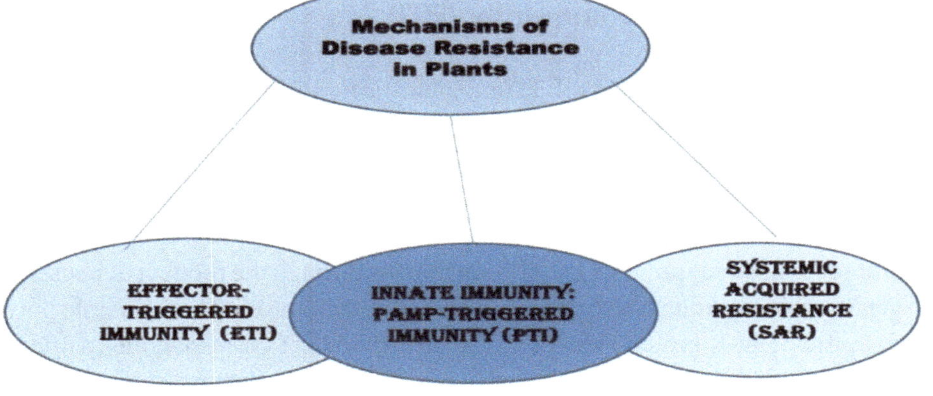

5.1 Innate Immunity: Pump Triggered Immunity (PTI)

Plants protect themselves against pathogens through PAMP-Triggered Immunity, a conserved innate immunity that is activated when detected by Pattern Recognition Receptors (PRRs) on the plant cell surface by conserved microbial molecules known as Pathogen-Associated Molecular Patterns (PAMPs). These PAMPs, like bacterial flagellin (recognized by utilization of the FLS2 receptor) or fungal chitin (recognized by utilization of CERK1), are vital for survival of pathogens and are thus the best targets for immune recognition. PRRs are typically transmembrane proteins like Receptor-Like Kinases (RLKs) having EFR (EFR, bacterial EF-Tu receptor) or Receptor-Like Proteins (RLPs) dimerizing with signal partners to become PRRs. Detection of PAMPs by PRRs triggers a signal cascade involving calcium entry, ROS bursts, and the activation of MAP kinases for transcriptional reprogramming. This results in physical defenses through the deposition of callose into cell walls, stomatal closure to close the path for pathogen entry, and the secretion of antimicrobial compounds like phytoalexins and pathogenesis-related (PR) proteins. While PTI induces broad-spectrum resistance, pathogens retaliate by effector protein release, which induces plants to activate a more potent, strain-specific defense known as Effector-Triggered Immunity (ETI) through intracellular NLR proteins. This dynamic interplay, described in the "Zig-Zag Model", highlights PTI's foundational role in non-host resistance and its agricultural relevance, as engineering PRRs (e.g., introducing EFR into wheat) has enhanced crop disease resistance.

5.2 Effector Triggered Immunity (ETI)

Effector-Triggered Immunity (ETI) is a sophisticated defense mechanism in plants that serves as the second layer of immune response also known as Gene-for-gene resistance (e.g., R genes), activated when pathogens circumvent the initial Pattern-Triggered Immunity (PTI). PTI relies on the recognition of conserved pathogen-associated molecular patterns (PAMPs), but successful pathogens often secrete effector proteins to suppress PTI. ETI counters this through the product of resistance (R) genes, which operate under the Gene-for-Gene hypothesis. This principle posits that specific plant R proteins directly or indirectly detect corresponding pathogen effector proteins (encoded by avirulence, Avr, genes), triggering a robust immune response. R proteins, typically nucleotide-binding leucine-rich repeat (NLR) receptors, may bind effectors directly or monitor host cellular components modified by effectors, a concept known as the "guard hypothesis." Recognition activates defense mechanisms, including a hypersensitive response (HR), characterized by localized programmed cell death to restrict pathogen spread, a reactive oxygen species burst, and the production of antimicrobial compounds. ETI also induces systemic acquired resistance (SAR), priming the entire plant for enhanced defense via salicylic acid signaling.The evolutionary arms race between plants and pathogens drives ETI dynamics: while R genes confer durable resistance in agriculture, pathogen effector mutations can evade detection, necessitating strategies like R gene stacking. Despite this, ETI remains critical for breeding disease-resistant crops, underscoring its role in both natural immunity and agricultural sustainability.

5.3. Systemic Acquired Resistance (SAR)

Systemic Acquired Resistance (SAR) is the plant's inducible defense strategy and offers long-term, broad-spectrum resistance to a variety of pathogens—like oomycetes, fungi, bacteria, and viruses—after the first localized infection or after exposure to some immune elicitors. SAR provides long-term protection with benefits such as less likelihood of pathogen evolution compared to R-gene resistance. It is a type of "trained immunity," utilizing systemic signaling for heightened future defence.The difference between PTI, ETI, and SAR is given below in Table 3.

Table 3. Difference between PTI ETI and SAR

Aspect	Pathogen-Triggered Immunity (PTI)	Effector-Triggered Immunity (ETI)	Systemic Acquired Resistance (SAR)
Trigger Mechanism	Identifies pathogen-associated molecular patterns (PAMPs)	Recognizes specific pathogen effectors delivered into plant cells	Induced after an initial infection, conferring long-term defense
Receptor Type	Pattern recognition receptors (PRRs)	Nucleotide-binding leucine-rich repeat (NLR) proteins	Salicylic acid-mediated signaling pathways
Response Speed	Immediate but generalized response	Stronger and more specific immune reaction	Gradual activation affecting the entire plant system
Defense Mechanism	Enhances cell wall integrity and initiates oxidative burst	Hypersensitive response (HR), leading to programmed cell death	Activates pathogenesis-related (PR) proteins, strengthening resistance
Pathogen Adaptation	Pathogens may evolve mechanisms to evade PRR detection	Pathogens can suppress ETI through effector modifications	Provides broad-spectrum resistance, limiting pathogen adaptation
Longevity of Immunity	Short-term and localized	Highly effective but temporary	Long-lasting and systemic
Key Molecules	Mitogen-activated protein kinases (MAPKs), callose deposition	Resistance (R) proteins, HR-associated molecules	Salicylic acid, PR proteins

5.4 Role of R-Genes and PR (Pathogenesis-Related) Proteins in Plant Immunity

Plants rely on a sophisticated two-tiered immune system to combat pathogens.

5.4.1 Resistance (R) genes)

Resistance (R) genes encode proteins that act as "sensors" to detect pathogens and initiate immune responses act as critical sentinels within the plant's innate immune system, providing highly specific protection against particular pathogen strains. Their primary role is to directly or indirectly detect the presence of specific pathogen virulence molecules, known as effectors (or avirulence factors, Avr factors), delivered by the invading microbe into the plant cell. These genes typically encode intracellular receptor proteins, often of the nucleotide-binding leucine-rich repeat (NB-LRR) class, which function as sophisticated surveillance molecules. This specific recognition event triggers a powerful defence response called as Effector-Triggered Immunity (ETI) which is characterized by a rapid, localized programmed cell death at the infection site, known as the hypersensitive response (HR), which effectively walls off the pathogen and limits its spread. Furthermore, R-gene activation initiates complex signaling cascades involving hormones like salicylic acid, leading to the establishment of Systemic Acquired Resistance.

5.4.2 PR Proteins (Pathogenesis-Related Proteins)

Pathogenesis-Related (PR) proteins are diverse group of plant-encoded proteins and an essential part of plant defense mechanisms, induced in response to pathogen attack (bacteria, fungi, viruses, nematodes), as well as other stresses like wounding, UV light, or chemical elicitors, which are activated after attack by the pathogen or exposure to stress. PR proteins play a vital role in Systemic Acquired Resistance (SAR) and Induced Resistance to protect plants from infection and increase immunity. Breakdown of the major families and their primary activities in Table 4.

Table 4. Breakdown of the major families and their primary activities

PR Family	Representative Members/Properties	Primary Activities & Roles in Defense
PR-1	"Antifungal" proteins; highly induced during SAR; unknown biochemical function	Antifungal activity; Marker for SAR; Possible role in sterol binding or membrane disruption?
PR-2	β-1,3-Glucanases	Hydrolyze β-1,3-glucans (major fungal cell wall component); Release defense elicitors; Antimicrobial
PR-3	Chitinases (Class I, II, IV, V, VI, VII)	Hydrolyze chitin (major fungal & insect cuticle component); Release defense elicitors; Direct antifungal
PR-4	Chitinases (Class I & II); Hevein-like; Win-like	Chitin-binding & hydrolysis (some members); Antifungal; Wound response
PR-5	Thaumatin-like; Osmotin; Zeamatin	Permeabilize fungal membranes (Osmotin-like); Antifungal; Anticomplement; Possibly hydrolyze β-1,3-glucans?
PR-6	Proteinase Inhibitors (e.g., Cysteine Protease Inhibitors)	Inhibit pathogen-derived proteases; Reduce nutrient availability; Deter herbivores
PR-7	Endoproteinases (e.g., P69 in tomato)	Unknown precise defense role; Possibly involved in processing other defense proteins
PR-8	Chitinases (Class III)	Hydrolyze chitin; Direct antifungal
PR-9	Peroxidases (Lignin-forming peroxidases)	Contribute to lignin/suberin formation (physical barrier); ROS production; Cell wall strengthening
PR-10	"Ribonuclease-like"; Bet v 1 homologs; Intracellular	Possible RNase activity (degrade pathogen RNA); Cytokinin binding; Abiotic stress tolerance
PR-11	Chitinases (Class V)	Hydrolyze chitin; Direct antifungal
PR-12	Defensins	Small, cysteine-rich; Permeabilize microbial membranes; Antifungal, antibacterial, insecticidal

continued on following page

Table 4. Continued

PR Family	Representative Members/Properties	Primary Activities & Roles in Defense
P-13	Thionins	Small, cysteine-rich; Highly toxic; Permeabilize microbial & plant membranes; Broad antimicrobial
PR-14	Lipid Transfer Proteins (LTPs)	Transfer phospholipids; Cutin synthesis; Permeabilize fungal membranes; Antifungal
PR-15	Oxalate Oxidase-like (Germin)	Produce H_2O_2; ROS signaling/toxicity; Cell wall cross-linking
PR-16	Oxalate Oxidase-like (Germin-like Prcteins - GLPs)	ROS signaling/toxicity; Cell wall strengthening; Some have superoxide dismutase activity
PR-17	Unknown enzymatic activity	Unknown function; Induced by pathogens

PR proteins function in plant immunity by following mechanisms:

- ***Direct antipathogen activity*:** Enzymatic degradation of pathogen cell walls, microbial growth inhibition, and membrane disruption of pathogens.
- ***Signalling and défense activation:*** Activating hypersensitive responses (HR) and systemic resistance against subsequent infections.
- ***Cell wall reinforcement:*** Making plant cell walls more resistant to pathogen penetration.

The hydrolytic activity of enzymes like PR-2 and PR-3 not only damages pathogens directly but also liberates molecular fragments (Pathogen-Associated Molecular Patterns - PAMPs, or Damage-Associated Molecular Patterns - DAMPs) from the damaged invaders. These fragments act as potent signals, triggering and intensifying broader plant immune responses. Additionally, PR proteins contribute to signalling through the generation of Reactive Oxygen Species (ROS). Enzymes such as peroxidases (PR-9), germin (PR-15), and germin-like proteins (PR-16) produce ROS, which function both as toxic agents against pathogens and as critical signalling molecules within the plant's defense network. The accumulation of specific PR proteins, notably PR-1, PR-2, and PR-5, serves as a well-established molecular indicator signalling the activation of long-lasting, systemic resistance (Systemic Acquired Resistance - SAR) throughout the plant.

5.5. Synergy Between R-Genes and PR Proteins

The overlapping function of R-genes and pathogenesis-related (PR) proteins establishes a multi-layered defense system within plants. R-genes are specialized guardians that recognize pathogen effectors at initial sites of infection, triggering hyper-sensitive responses (HR) that rapidly isolate invaders at localized infection sites by programmed cell death. At the same time, this recognition triggers salicylic acid (SA)-dependent signaling cascades that systemically amplify the secondary plant defenses.

6. GENETIC ENGINEERING TOOLS AND TECHNIQUES

6.1 Transgenic Approaches

It involve introduction of foreign genes into plants to confer resistance against pathogens. These methods can leverage advancements in molecular biology to enhance plant immunity, reduce crop losses, and minimize reliance on chemical pesticides. Below are the key strategies, applications, and challenges in this field, supported by evidence from recent research.

6.1.1 Pathogen-derived resistance (PDR)

This involves introduction of genes from the pathogen itself into the plant genome. Transgenic expression of viral coat proteins in squash and papaya confers resistance to viruses such as *Papaya ringspot virus* (PRSV) by triggering RNA interference (RNAi). Similarly, potato plants engineered with viral replicase genes show resistance to *Potato leafroll virus* (Fuchs & Gonsalves, 2007).

6.1.2. Antimicrobial peptides (AMPS) and toxins

This involves heterologous expression of genes encoding antimicrobial compounds enhances the resistance. The Cecropin, a synthetic moth-derived AMP, confers resistance to fungal (*Fusarium*) and bacterial pathogens in rice and citrus (Buziashvili et al, 2023).

6.1.3. Resistance (r) genes and nlr proteins

The nucleotide-binding leucine-rich repeat (NLR) proteins detect pathogen effectors and activate immune responses. The *Bs2* gene from pepper transferred to tomato confers resistance to bacterial spot disease (Lee et al, 2016). Similarly, *RGA2* from wild bananas enhances resistance to *Fusarium* in Cavendish bananas (Dale et al, 2017). Four genetically modified (GM) TR4-resistant Cavendish (cv. Grand Nain) banana lines (RGA2-2, -3, -4, and -5), each transformed with the *MamRGA2* resistance gene derived from a TR4-resistant wild banana (Harding et al.2025).

6.1.4. Pattern recognition receptors

The pattern recognition receptors (PRRs) from resistant species enhances pathogen detection. Transferring *EFR* (Elongation Factor–Tu Receptor) from Arabidopsis to tomato and wheat improves resistance to bacterial pathogens like *Ralstonia solanacearum* (Cook et al, 2015). The rice receptor kinase *Xa21* confers resistance to *Xanthomonas* wilt in bananas (Buziashvili et al, 2023).

6.1.5. RNA interference (RNAi)

The RNAi silences pathogen genes by expressing double-stranded RNA (dsRNA). Transgenic papaya expressing PRSV coat protein-derived RNAi shows durable resistance (Fuchs & Gonsalves, 2007). RNAi targeting fungal chitin synthases or insect digestive enzymes (e.g., *Snf7*) reduces pathogen viability (Buziashvili et al, 2023).

6.1.6. Gene stacking and genome editing

Combining multiple "R" genes or CRISPR-edited susceptibility (S) genes (e.g., *mlo* in wheat) provides broad-spectrum protection (Ali et al, 2022)

6.2 Cisgenic and Intragenic Strategies

Cisgenesis involves the transfer of a complete gene (including its native promoter, introns, and terminator) from a sexually compatible donor plant to a recipient plant. This approach avoids linkage drag by introducing only the desired gene. For example, the *Rpi-blb2* gene from wild potato (*Solanum* bulbocastanum) was cisgenically introduced into cultivated potato to confer resistance to *Phytophthora infestans* (late blight) (Haverkort et al, 2016).

Intragenesis allows recombination of genetic elements (e.g., promoters, coding sequences) within the same species, enabling synthetic rearrangements such as antisense orientation or RNA interference (RNAi) cassettes. Unlike cisgenesis, intragenesis permits modifications like chimeric gene constructs but restricts DNA sources to the same or sexually compatible species (Rommens et al, 2007). Strawberries have been engineered with RNA interference (RNAi) constructs targeting fungal chitin synthase (CHS) genes, effectively lowering their vulnerability to *Fusarium oxysporum* infections (Schaart et al., 2016).

6.3. Genome editing for disease resistance in plants

Current research actively explores the practical applications and inherent limitations of following gene editing technique within agricultural genetic engineering.

6.3.1 CRISPR-CAS9

The most widely adopted system, utilizing a guide RNA to direct the Cas9 nuclease to specific DNA sequences. It creates double-strand breaks (DSBs) at targeted genomic locations, which the cell's repair mechanisms then fix, often leading to gene knockouts or enabling precise insertions if a repair template is provided. The targeted knockout of the *TaMLO* locus in wheat chromosomes confers durable resistance to powdery mildew by preventing fungal penetration, while inactivation of the SlDMR6-1 gene in tomato enhances innate immunity against *Pseudomonas syringae* by elevating salicylic acid accumulation. The *mlo* genes, which encode membrane proteins critical for powdery mildew susceptibility, have been targeted in wheat, tomato, and barley to confer durable resistance (Sánchez-Martín et al. 2021; Wang et al, 2025).

6.3.2 CRISPR-CAS12 (CPF1)

Beyond standard Cas9, engineered variant systems significantly expand application horizons to Cas12 (Cpf1), derived from *Prevotella* and *Francisella* bacteria, recognizes T-rich PAM sites (TTTV) and generates staggered DNA cuts with 5′ overhangs. This cleavage profile enhances homology-directed repair (HDR) efficiency when inserting resistance alleles, proving particularly valuable for editing promoter regions of *OsSWEET* genes in rice to combat bacterial blight. Cas12 (Cpf1) represents a distinct class of CRISPR-associated nucleases with structural and functional features that expand genome-editing capabilities beyond Cas9. Unlike Cas9's blunt-ended cuts, Cas12a employs a single RuvC domain to generate staggered double-strand breaks with 5′ overhangs, enhancing homology-directed repair (HDR) efficiency for precise allele insertion. For example, In rice, Cas12a-mediated editing within OsSWEET11/14 genes disrupts *Xanthomonas oryzae* virulence and develop bacterial blight immunity.

6.3.3. CRISPR-Cas13

This is primarily used for transient RNA knockdown (degradation) or potentially RNA base editing, offering a way to modulate gene expression without altering the underlying DNA genome. These evolutionary branches of CRISPR technology collectively address critical limitations: Cas12 broadens targetable genomic sites in AT-rich regions where NGG PAMs are scarce, while Cas13 enables transient therapeutic effects ideal for combating rapidly evolving RNA viruses. Their ongoing refinement continues to unlock novel disease management strategies across global cropping systems.

6.3.4. Base editing

It revolutionizes the engineering of disease-resistant crops by enabling precise single-nucleotide conversions without inducing double-strand DNA breaks (DSBs) or requiring exogenous donor templates. This technology fuses catalytically impaired Cas9 variants (typically D10A nickase mutants) with nucleotide-modifying enzymes, cytidine deaminases for C•G→T•A conversions or evolved adenosine deaminases for A•T→G•C transitions. The molecular architecture functions through localized DNA unwinding within a ~5-nucleotide activity window, where the deaminase domain chemically alters target bases before cellular repair mechanisms finalize permanent base substitutions. This precision is particularly valuable for introducing gain-of-function mutations in resistance (R) genes or disrupting susceptibility factors through targeted point mutations. For example, converting a specific cytosine to thymine in the rice *OsERF922* promoter region creates a premature stop codon that represses this negative regulator of immunity, resulting in enhanced resistance to *Magnaporthe oryzae* (rice blast fungus) without collateral genomic damage. Similarly, adenine-to-guanine transitions in wheat *TaMLO* homologs precisely recreate natural loss-of-function alleles that confer powdery mildew resistance while avoiding the genomic rearrangements associated with DSB-dependent editing. The technology's avoidance of DSBs proves critical for polyploid crops like potato, where conventional editing often causes chromosomal instability base editing of the DMR6-1 susceptibility gene against *Phytophthora infestans* maintains genomic integrity while achieving pathogen resistance. Recent advances include dual base editors for simultaneous C→T and A→G conversions in maize *ZmIPK* genes, broadening resistance spectra against multiple fungal pathogens through coordinated amino acid substitutions in defense-related proteins.

6.3.5 TALENs (Transcription Activator-Like Effector Nucleases)

TALENs represent a class of programmable nucleases engineered through the fusion of two functional domains: a customizable DNA-binding array derived from *Xanthomonas* bacterial TAL effectors, and the catalytic domain of the *FokI* endonuclease (Li et al. 2012). The DNA-binding component comprises tandem repeats of 33–35 amino acid modules, where hypervariable residues at positions 12 and 13 (known as

Repeat Variable Di-residues or RVDs) dictate nucleotide specificity—NI for adenine, HD for cytosine, NG for thymine, and NN for guanine. This modular architecture enables precise targeting of virtually any genomic sequence by assembling RVD arrays in a defined order. For gene editing, paired TALEN monomers bind opposing DNA strands at spacings of 12–20 base pairs, facilitating dimerization of the *FokI* domains. This dimerization triggers double-strand DNA cleavage, generating cohesive ends with 5′ overhangs that engage cellular repair pathways. Unlike CRISPR-Cas systems, TALENs function independently of protospacer adjacent motifs (PAMs), granting unique access to GC-rich genomic regions often recalcitrant to Cas9 editing. In plant pathology, this capability has been leveraged to disrupt bacterial blight susceptibility genes; for example, editing the effector binding elements (EBEs) in the promoter of rice *OsSWEET14* abolishes *Xanthomonas oryzae* virulence by preventing transcriptional activation of this sucrose transporter. Similarly, TALEN-mediated knockout of cassava *MeSWEET10a* confers resistance to *Xanthomonas axonopodis* without yield penalties. The technology's high specificity—attributed to the requirement for dual monomer binding—minimizes off-target effects, though its adoption has been tempered by complex plasmid assembly protocols and reduced efficiency in polyploid crops. Recent innovations like Platinum TALENs with enhanced cleavage activity and streamlined Golden Gate assembly now enable multiplexed editing of stacked resistance genes, as demonstrated in tomato for concurrent defense against *Pseudomonas syringae* and *Fusarium oxysporum*

6.3.6 ZFNS (zinc finger nucleases)

Among the first programmable genome editors, ZFNs link engineered zinc finger protein domains (each recognizing a short DNA triplet) to the FokI nuclease. Similar to TALENs, ZFN pairs dimerize at the target site to create a DSB. While effective, designing highly specific zinc finger arrays can be more challenging than designing CRISPR guide RNAs. ZFNs represent the first generation of programmable genome editors, predating CRISPR systems, and remain valuable for specific agricultural applications. The zinc finger scaffold employs a conserved ββα topology where solvent-facing residues at positions −1, +2, +3, and +6 determine nucleotide specificity through modular combinatorial assembly. Functional editing requires paired ZFN monomers binding to sense and antisense DNA strands with a 5–7 bp spacer between target half-sites. Dimerization of the *FokI* domains induces double-strand breaks (DSBs) with 5′ overhangs, triggering endogenous repair pathways that facilitate targeted mutagenesis or gene insertion. Unlike PAM-dependent CRISPR systems, ZFNs access GC-rich genomic regions prevalent in plant pathogen susceptibility genes. This capability enabled foundational breakthroughs in crop disease resistance, including, Disruption of rice *Xa13* promoter elements to create bacterial blight-resistant lines by abolishing *Xanthomonas oryzae* effector binding. Knock-in of mildew resistance locus (*MLO*) mutations in barley through homology-directed repair (HDR), replicating natural loss-of-function alleles that block powdery mildew penetration. Multiplexed editing of wheat *Pm3* alleles to broaden fungal resistance spectra without linkage drag.

Despite these successes, ZFN adoption faces challenges: context-dependent finger specificity necessitates laborious empirical optimization, and off-target cleavage may occur at sites with >70% homology. Recent advances like OPEN-ZFNs (Oligomerized Pool ENgineering) and CoDA (Context-Dependent Assembly) platforms streamline design while high-fidelity *FokI* variants (e.g., Sharkey mutations) reduce dimerization-independent activity. In cassava, optimized ZFNs achieved 12% HDR efficiency in *MeSWEET10a*—a bacterial susceptibility gene—demonstrating renewed potential for editing crops where CRISPR faces delivery limitations .

6.3.7 Meganucleases (homing endonucleases):-

Meganucleases, also termed homing endonucleases, represent a distinct class of naturally occurring DNA-cutting enzymes originally discovered within mobile genetic elements of microbes like yeast, algae, and bacteria. What sets these proteins apart is their exceptional ability to recognize and bind extremely long DNA sequences—typically spanning 12 to 40 base pairs—a feature that grants them unparalleled intrinsic specificity. This precision arises from their complex structural features, including conserved catalytic motifs such as LAGLIDADG, which enable both DNA binding and cleavage in a single integrated domain. In their natural settings, these enzymes facilitate their own propagation by creating targeted double-strand breaks in genomes lacking the meganuclease gene, hijacking the cell's repair machinery to copy themselves into new sites—a process termed "homing."

While their natural precision is remarkable, wild-type meganucleases rarely target sequences useful for biotech applications. This limitation spurred efforts to engineer custom versions through advanced protein redesign methods. Scientists employ computational modeling, directed evolution, and domain swapping to retool these enzymes, enabling them to cut predetermined genomic locations. For instance, engineered variants of enzymes like I-CreI have been reprogrammed to disrupt human disease-related genes or modify crop genomes. Yet despite this capability, meganucleases face practical constraints that restrict their widespread use. Creating each new variant demands extensive expertise and laborious optimization, lacking the straightforward, modular design of newer tools like CRISPR-Cas9 or TALENs. Their large size also complicates delivery via viral vectors for therapeutic applications.

Consequently, while engineered meganucleases demonstrated early promise in clinical trials (e.g., targeting the CCR5 gene for HIV resistance) and agriculture (e.g., developing herbicide-tolerant maize), they have largely been superseded by more versatile systems. Today, they occupy niche roles where extreme specificity outweighs development hurdles—such as editing safety-critical genomic sites where off-target effects could prove catastrophic. Hybrid approaches, like fusing meganuclease domains to TALE proteins, aim to revive their relevance by merging their precision with more flexible targeting. Thus, while no longer mainstream, meganucleases remain a testament to the evolution of gene-editing technology and continue to inform specialized applications demanding near-absolute accuracy.

7. CASE STUDIES OF GENETICALLY ENGINEERED DISEASE-RESISTANT CROPS

7.1. Virus-Resistant Papaya

The development of Rainbow Papaya (a genetically engineered virus-resistant papaya) during 1998, stands as a landmark achievement in crop genetic engineering. Dr. Dennis Gonsalves and their team that introduced the viral coat protein gene into papaya plants using genetic engineering which successfully saved the Hawaii's papaya industry and papaya cultivating farmers loss preventing infection and ensuring stable production. This innovation highlights the potential of genetic engineering in protecting crops, ensuring food security, and promoting sustainable agriculture.

7.2 Late Blight-Resistant Potato (RNAi and r-gene approaches)

To combat the Late blight (caused by *Phytophthora infestans)*, a major threat to global potato production often leading to severe crop losses, researchers have developed genetically engineered potatoes resistant to late blight using RNA interference (RNAi) and resistance (R) gene approaches (Sun et al, 2016; Mahfouze& El-Sayed, 2022). The RNAi strategy involves silencing susceptibility (*S*) genes that the pathogen exploits for infection, thereby preventing disease progression (Sun et al, 2016). Alternatively, the R-gene approach incorporates resistance genes such as Rpi-sto1 and Rpi-vnt1.1 from wild potato species, enabling the plants to recognize and block pathogen attack (Jo et al, 2014). These advancements have significantly reduced the need for chemical fungicides, promoting more sustainable and environmentally friendly potato cultivation (Mahfouze& El-Sayed, 2022). By integrating these technologies, scientists aim to enhance crop resilience, safeguard food security, and provide farmers with disease-resistant varieties that require fewer agrochemical inputs. The success of genetically engineered late blight-resistant potatoes highlights the potential of biotechnology in modern agriculture, paving the way for improved plant disease management strategies.

7.3 Fusarium-Resistant Wheat

Genetically engineered Fusarium-resistant wheat has emerged as a promising solution to combat *Fusarium graminearum*, the causal agent of Fusarium head blight (FHB), which leads to severe yield losses and mycotoxin contamination. Traditional breeding methods have struggled to provide durable resistance, but advances in genetic engineering, CRISPR-based modifications, and multi-omics approaches have significantly improved wheat's defense mechanisms (Ma et al, 2025; Wang et al, 2025). One strategy involves CRISPR/Cas9-mediated gene editing, where susceptibility (*S*) genes such as TaHRC and Tsn1 are knocked out to reduce fungal colonization and toxin accumulation (Wang et al, 2025). Additionally, researchers have introduced Fhb1 and Fhb7, well-characterized resistance genes, using transgenic approaches to strengthen wheat's natural immunity (Ji et al, 2025). Comparative transcriptomic studies have revealed unique defense mechanisms in resistant wheat cultivars, including the upregulation of genes involved in pathogen destruction and toxin detoxification (Ji et al, 2025). Genome-wide association studies (GWAS) have further identified significant marker-trait associations (MTAs) linked to FHB resistance, aiding in the development of more resilient wheat varieties (Ma et al, 2025). These advancements have significantly enhanced wheat resilience, reducing dependence on fungicides and promoting sustainable agricultural practices. The integration of genomics, transcriptomics, and metabolomics has refined resistance mechanisms, paving the way for next-generation Fusarium-resistant wheat varieties.

7.4 EIF4E Gene Editing

The eIF4E gene plays a crucial role in plant-virus interactions, serving as a susceptibility factor that viruses exploit for replication. In efforts to enhance viral resistance, researchers have employed CRISPR/Cas9-mediated genome editing to modify eIF4E in crops such as cucumber and potato (Noureen et al, 2022). In potato, studies have demonstrated that disrupting the interaction between eIF4E and the viral VPg protein significantly reduces susceptibility to Potato Virus Y (PVY) (Noureen et al, 2022). CRISPR/Cas9-mediated mutations in conserved regions of eIF4E resulted in truncated proteins that failed to support viral replication, leading to enhanced resistance in tetraploid potato lines. Similarly, in cucumber, targeted genome editing of eIF4E has been shown to confer resistance against multiple viruses, including Cucumber Mosaic Virus (CMV) and Zucchini Yellow Mosaic Virus (ZYMV). These advancements highlight the potential of precision breeding in developing virus-resistant crops, reducing reliance on chemical treatments, and promoting sustainable agriculture. The integration of non-transgenic genome editing approaches further enhances the feasibility of deploying virus-resistant cultivars in commercial agriculture .

7.5 Bt Cotton

Bt cotton is a genetically engineered crop that incorporates *Bacillus thuringiensis* (Bt) genes to provide resistance against bollworm pests, reducing reliance on chemical insecticides. Using techniques such as Agrobacterium-mediated transformation and gene gun methods, scientists introduce cry genes into cotton plants, enabling them to produce insecticidal proteins that target lepidopteran larvae. These proteins bind to receptors in the pests' gut, causing cell lysis and ultimately leading to their death. Bt cotton has significantly improved crop yields, minimized environmental pesticide pollution, and lowered production costs for farmers by reducing pest damage. However, continuous planting of Bt cotton requires resistance management strategies, such as refuge crops and gene stacking, to prevent pest adaptation and ensure long-term effectiveness.

Bt crops, genetically modified to express *Bacillus thuringiensis* (Bt) toxins, provide built-in insect resistance, reducing dependence on chemical pesticides and enhancing crop yields (Abbas, 2018; Selvam &Ragiman, 2024). These crops, including Bt cotton, Bt corn, and Bt brinjal, effectively target major agricultural pests such as lepidopteran and coleopteran insects, minimizing plant damage and promoting healthy growth (Abbas, 2018). Beyond their primary function of insect resistance, Bt crops offer secondary disease benefits, as reduced pest infestations limit wounds that serve as entry points for bacterial and fungal pathogens (Pan, 2024). For instance, Bt corn has demonstrated decreased susceptibility to Fusarium ear rot, a fungal disease that produces harmful mycotoxins, due to reduced insect damage (Abbas, 2018). Similarly, Bt cotton exhibits lower occurrences of bacterial and fungal infections, as bollworm resistance prevents wounds that facilitate pathogen invasion (Selvam &Ragiman, 2024). In Bt brinjal, resistance to fruit borers has led to a significant decline in fruit rot and bacterial wilt, ensuring better crop quality (Pan, 2024). These indirect disease protection benefits contribute to sustainable pest management, reducing agricultural losses and improving food security. The integration of Bt technology has revolutionized pest control strategies, offering environmentally friendly solutions that minimize chemical inputs while while fostering resilient crop production.

7.6. Powdery Mildew Resistant Wheat

One of the recent breakthroughs in genome editing for plant disease resistance involves the targeted disruption of the *MLO* gene in wheat using CRISPR-Cas9 technology. Powdery mildew, caused by *Blumeria graminis* f. sp. *tritici*, is a widespread fungal disease that significantly reduces wheat productivity and grain quality (Li et al, 2021).

7.7. Tomato With Resistance to Gemini Viruses

The use of CRISPR-Cas9 genome editing to enhance tomato resistance against geminiviruses, particularly the Tomato leaf curl New Delhi virus (ToLCNDV targeting the viral genes *AC2* and *AC3*, which are essential for viral replication and suppression of host defense mechanisms. The success of this strategy highlights the potential of genome editing in developing virus-resistant cultivars (Wang et al., 2025).

8. CHALLENGES IN GENETIC ENGINEERING FOR DISEASE RESISTANCE IN PLANTS

8.1. Pathogen Evolution and Resistance Breakdown

Pathogens rapidly evolve to overcome resistance genes which leading to the breakdown of genetic resistance for diseases in plants. For example, wheat stem rust (*Puccinia graminis*) mutated its AvrSr31 effector to bypass the Sr31 resistance gene, leading to the devastating Ug99 strain that compromised global wheat production.

8.2. Off-Target Effects in Genome Editing

One of the most important concerns in genome editing is the occurrence of off-target effects, particularly when utilizing CRISPR-based technologies. Such modifications can inadvertently disrupt important regulatory genes or coding sequences unrelated to the trait being edited. As a result, unpredictable and unwanted phenotypes develop. Therefore, rigorous validation strategies and advanced guide RNA design tools are very essential to minimize such risks in genome-edited crops,

8.3. Ecological Risks And Gene Flow

One of the critical concerns associated with genome editing in agriculture is the potential ecological impact stemming from unintended gene flow. When genetically modified or genome-edited crops are cultivated in open environments, there is a possibility that novel genes may escape into wild or weedy relatives through natural mechanisms such as cross-pollination. This unintended gene transfer can lead to the expression of traits such as herbicide resistance or altered growth characteristics in non-target plant populations. Over time, these changes may disrupt local ecosystems by giving certain plants a selective advantage, which could shift species dynamics or even displace native flora. Furthermore, the resulting hybridization may introduce genetic instability, complicating conservation efforts and threatening long-standing biodiversity. As emphasized by Dong et al. (2019), such risks underscore the importance of implementing robust containment strategies and ecological monitoring to safeguard the balance between agricultural advancement and environmental sustainability.

8.4. Public Perception and GMO Acceptance

Public perception plays a pivotal role in determining the success and acceptance of genetically modified organisms (GMOs), including those developed through genome editing technologies. Despite scientific advancements and regulatory assurances, consumer skepticism remains a significant barrier, often fueled by concerns about food safety, ethical implications, and environmental impact. This hesitancy affects not only market adoption but also policy decisions and research investments. Misinformation and lack of transparency can further deepen public distrust, making it crucial for scientists, regulatory authorities, and industry stakeholders to engage in clear, fact-based communication. Building trust through accessible education, participatory decision-making, and rigorous safety evaluations can bridge the gap between scientific innovation and societal acceptance. As highlighted by Sindhu et al. (2023), fostering informed dialogue and ensuring regulatory clarity are key to achieve broader acceptance of genome-edited crops.

8.5. Regulatory and Biosafety Challenges

The implementation of genome editing in agriculture is often hindered by complex regulatory frameworks and stringent biosafety protocols that vary significantly across countries. These regulations are designed to ensure environmental and food safety, but the absence of globally harmonized standards can lead to prolonged and unpredictable approval timelines for genome-edited crops. Such delays not only restrict the pace of innovation but also impede timely market entry and commercialization. Additionally, inconsistent regulatory classifications—where some nations treat genome-edited organisms similarly to traditional GMOs while others exempt them under specific conditions—create challenges for international trade and collaboration. The resulting uncertainty discourages investment, particularly from smaller enterprises and public sector institutions. As noted by Dong et al. (2019), balancing safety assessments with regulatory efficiency is essential to support innovation while maintaining public and environmental confidence in emerging biotechnologies.

8.6. Limited Resistance Spectrum

A notable limitation in genome-engineered disease resistance is its often narrow or strain-specific efficacy. While biotechnological tools such as CRISPR have enabled the development of crops with targeted resistance traits, these enhancements are frequently tailored to specific pathogen strains or biotypes. This means that while a crop may exhibit strong defense against a known variant, it might remain susceptible to other evolving or uncharacterized strains of the same pathogen. Pathogens, especially fungi, bacteria, and viruses, are known to adapt rapidly, and this evolutionary flexibility can render engineered resistance partially or wholly ineffective over time. Consequently, reliance on such targeted resistance could lead to unexpected disease outbreaks, crop losses, or increased use of agrochemicals for management. As Sindhu et al. (2023) emphasized, achieving durable and broad-spectrum resistance remains a central challenge, necessitating integrated strategies that combine genetic engineering with traditional breeding and holistic plant health practices

8.7. High Development Costs

The high financial burden associated with the research and development (R&D) of genetically engineered crops poses a significant obstacle to their widespread adoption, particularly among smallholder farmers. The development process involves costly components such as advanced laboratory infrastructure, skilled personnel, regulatory compliance, and extensive field trials—all of which contribute to elevated production costs. These investments are often justifiable for large agribusinesses aiming at high-value markets, but they can be prohibitively expensive for resource-constrained farmers in developing regions. Moreover, the need to recoup R&D investments can lead to increased seed prices or licensing fees, further limiting affordability. This financial disparity creates a technological divide where the benefits of modern genetic innovations remain concentrated among industrial-scale operations. As noted by Dong et al. (2019), fostering inclusive access to agricultural biotechnology requires innovative policy interventions, public-private partnerships, and support mechanisms tailored to the needs of small-scale farming communities.

9. FUTURE DIRECTIONS IN PLANT DISEASE RESISTANCE THROUGH GENETIC ENGINEERING

Unlike earlier tools, high-fidelity variants like Cas12b exploit AT-rich genomic regions—editing promoter sequences of susceptibility genes that traditional nucleases could not access. Prime editing further revolutionizes precision by facilitating single-nucleotide substitutions without DNA breaks, exemplified in rice where *Os-SWEET14* edits confer robust resistance against *Xanthomonas oryzae* by disrupting pathogen nutrient access. Similarly, base editors introduce targeted C→T conversions in key loci; wheat and barley edited at *mlo* genes exhibit near-complete powdery mildew immunity by mimicking natural loss-of-function mutations. These advances empower two strategic applications of susceptibility gene knockout, as seen in banana *MaDMR6* inactivation that reduces Fusarium TR4 infection by >70%, and regulatory element engineering, where citrus *LOB1* promoter modifications block *Xanthomonas* effector binding. Although there is outstanding progress in genetic engineering for disease resistance then also there are some gaps in the knowledge to be explored.

9.1. Long Term Resistance Efficacy

While gene stacking confers resistance, long-term efficacy of stacked genes against evolving pathogens is not yet proven and requires further investigation on pathogen adaptation mechanisms (Dong et al, 2019).Term Evolutionary Durability of Stacked Resistance GenesA critical unresolved challenge in genetically engineered disease resistance concerns the uncertain long-term stability of gene stacking strategies against rapidly co-adapting pathogens. While pyramiding multiple resistance genes (e.g., CRISPR-edited *R* genes, PRRs, and susceptibility factor knockouts) demonstrates robust initial efficacy in controlled trials, the *evolutionary trajectories* of pathogen populations under sustained selective pressure remain poorly characterized. Field validations rarely exceed 3-5 growing seasons—insufficient to capture compensatory virulence mechanisms like effector gene recombination or epistatic mutation networks that restore pathogen fitness. For instance, the collapse of *Pi2+Piz5* stacked resistance in rice within 8 years, where *Magnaporthe oryzae* evolved novel *Avr-Pik* haplotypes through genomic rearrangements undetectable in short-term studies. This gap is compounded by inadequate modeling of abiotic stressors (e.g., heat waves accelerating mutation rates) and incomplete fitness-cost analyses of virulence alleles across diverse agro-ecosystems. Current deployment frameworks lack predictive algorithms for resistance breakdown thresholds, leaving breeding programs vulnerable to unanticipated failure. Without longitudinal pathogen surveillance integrated with climate-resilience phenotyping, the durability of multiplexed genetic solutions remains precariously speculative

9.2. Unmitigated Off-Target Mutational Consequences in CRISPR-Edited Disease Resistance

Though CRISPR/Cas9 is more precise, off-target mutations in non-target loci can influence plant growth and development, and high-fidelity genome editing methods become the need of the hour.The precision limitations of CRISPR-Cas9 systems present a critical barrier to engineering durable disease resistance in crops, where even minimal off-target activity—often undetectable in standard screenings—induces deleterious pleiotropic effects that compromise plant fitness. Off-target gRNA-Cas9 complexes frequently bind to chromatin-topologically associated domains (TADs), inadvertently silencing defense-related genes like "PR-1" through aberrant methylation shifts or triggering chromosomal rearrangements within paralogous NLR gene clusters. More insidiously, frameshift mutations in metabolic regulators (e.g., phenylpropanoid pathway enhancers) can cripple phytoalexin biosynthesis, leaving plants vulnerable to secondary infections. Current mitigation approaches remain fundamentally inadequate: high-fidelity Cas variants like SpCas9-HF1 sacrifice substantial on-target efficiency in monocots, while chemical inhibitors of non-homologous end joining disrupt beneficial somatic recombination. Bioinformatic prediction tools further fail to account for plant-specific chromatin dynamics, as evidenced by Sindhu et al (2023) documenting 137 unintended mutations in *Pm3*-edited wheat lines—including a 4-bp deletion in *TaWOX5* that imposed 40% yield penalties and chromothripsis events disrupting *TaBAK1*-mediated immunity. Compounding these issues, no delivery system achieves cell-type-specific editing without systemic off-target effects, and stress-induced chromatin reconfigurations remain uncharacterized for most crops. Promising solutions like sgRNA-cDNA heteroduplexing or nanoparticle-conjugated Cas9-RNPs with tissue-specific targeting require rigorous validation to address this multidimensional challenge.

9.3. Fragmented Multi-Omics Integration In Disease Resistance Engineering

Use of genomics, transcriptomics, proteomics, and metabolomics for refinement of the disease resistance mechanism continues to be insufficient, with data-driven predictive models being the need (Dong et al, 2019).A critical methodological deficiency persists in deploying genomics, transcriptomics, proteomics, and metabolomics for refining plant disease resistance: the absence of unified predictive frameworks capable of capturing dynamic inter-omics crosstalk during host-pathogen interactions. Current approaches operate in disciplinary silos—genomics identifies resistance gene candidates, transcriptomics profiles defense gene induction, proteomics catalogs post-translational modifications, and metabolomics maps biochemical shifts—yet these layers remain disconnected rather than integrated. This fragmentation obscures essential mechanistic relationships, such as how SNP variations in NLR immune receptors (*genomic layer*) regulate phosphorylation-mediated signal transduction cascades (*proteomic layer*), which subsequently modulate phenylpropanoid flux (*metabolomic layer*) during fungal challenge. Compounding this gap, technical heterogeneity across platforms (e.g., RNA-seq depth versus mass spectrometry sensitivity) introduces normalization artifacts that mask causal links between molecular tiers. For instance, transcriptomic upregulation of PR-1 often poorly correlates with protein abundance due to miRNA-mediated silencing—a disconnect unquantified in most resistance studies. Consequently, machine learning models capable of predicting outcomes like CRISPR-edited *R* gene durability or pathogen effector evolution lack training datasets that simultaneously incorporate allelic diversity, stress-responsive proteoforms, and defense metabolite kinetics. Without computational architectures that resolve spatiotemporal dependencies across omics dimensions—such as digital twins simulating infection timelines or Bayesian networks correlating SNP haplotypes with metabolite covariance—breeding programs remain confined to reductionist strategies vulnerable to rapid pathogen adaptation, as evidenced by Dong et al (2019) finding that unmodeled metabolic bottlenecks accelerated resistance erosion in gene-stacked rice.

9.4. Ecological And Environmental Impact

Gene flow of introduced resistance genes to the wild relatives and added value to biodiversity necessitate long-term ecological evaluations. A critical deficit persists in evaluating the long-term agroecological impacts of genetically engineered disease resistance, particularly concerning uncontrolled transgene introgression into wild relatives and its cascading effects on biodiversity equilibrium. Current risk assessments prioritize short-term agronomic performance over multi-trophic consequences, failing to model how resistance genes (e.g., CRISPR-edited NLR alleles) transferred to wild populations via pollen-mediated gene flow may disrupt co-evolved pathogen-host dynamics in natural ecosystems. For instance, constitutive *Bs2* expression in wild *Solanum* spp. could suppress native pathogen biodiversity that sustains endemic arthropod food webs, inadvertently triggering trophic cascades. Compounding this gap is the unaddressed pleiotropy of resistance transgenes: engineered traits like chitinase overexpression may alter rhizosphere microbiomes or root exudate profiles, restructuring soil communities that regulate nutrient cycling—effects rarely monitored beyond single growing seasons. Sindhu et al (2023) documented such unintended outcomes in wheat-barley hybrid zones, where *Pm3* transgene escape reduced mycorrhizal colonization by 40% and suppressed beneficial *Pseudomonas* spp. essential for nitrogen fixation.

9.5 Unresolved to Socio-Regulatory Barriers Engineered Disease Resistance Commercialization

The biosafety issues and GMO acceptance hurdles still restrain commercialization, demanding open risk assessment and more robust regulation systems (Dong et al, 2019).Persistent biosafety concerns and entrenched public skepticism toward genetically modified crops continue to impede the translation of engineered disease resistance from research to field deployment, revealing critical deficiencies in regulatory science and stakeholder engagement frameworks. Current governance systems operate through fragmented, nationally divergent protocols that lack harmonized risk-assessment criteria—particularly for emerging technologies like transgene-free CRISPR edits—creating market access barriers that disproportionately disadvantage public-sector innovations. This socio-technical incongruence manifests in two interrelated failures: scientifically, most regulations prioritize *ex ante* lab-based hazard identification (e.g., allergenicity screens) while neglecting longitudinal agroecological monitoring of gene flow impacts on soil microbiome resilience or non-target arthropod communities; societally, top-down communication strategies disregard cultural cognition variables that shape risk perception, such as institutional trust deficits among smallholders or ethical objections to "unnatural" genome interventions. Consequently, even empirically low-risk traits like mildew-resistant CRISPR-Cas9-free barley face activist litigation and import bans. Documentation of 17 abandoned field trials despite peer-reviewed safety data. Compounding this barrier is the inadequate participatory governance: Indigenous knowledge systems and farmer experiential data remain excluded from regulatory decision matrices, while opaque approval processes fuel conspiracy narratives. Resolving this impasse demands co-designed frameworks integrating 1) *adaptive biosafety metrics* quantifying ecological redundancy loss from stacked-gene introgression, 2) *blockchain-tracked stewardship* protocols enabling consumer verification, and 3) *deliberative democracy models* embedding citizen juries in technology assessment. Without such multidimensional reforms, the $3B annual R&D investment in engineered disease resistance risks generating stranded innovations with negligible societal benefit.

9.6. Climate-Resilient Disease Resistance

The interaction between the climate stress factors (drought, heat) and disease resistance genes is not well elucidated and requires research on climate-adaptive genetic alterations. A critical deficit exists in understanding how climate stressors—particularly drought and heat extremes—disrupt molecular defense mechanisms in genetically engineered crops, creating unintended pathogen adaptation bottlenecks that accelerate resistance erosion. Current breeding paradigms assume static gene function across environments, yet temperatures exceeding 32°C induce thermolabile misfolding of NLR immune receptor domains (e.g., NB-ARC structural denaturation in *Bs2*), while drought stress impairs RNAi-mediated viral defense through Dicer-like protein dysregulation. These abiotic perturbations concurrently select for climate-adapted pathogen strains: *Fusarium graminearum* evolves heat-stabilized Hsp90 chaperones to enhance effector delivery precisely when host immunity weakens, as Sindhu et al (2023) demonstrated in wheat where CRISPR-edited *Pm3a* resistance collapsed under combined heat-drought stress due to three synergistic failures—disrupted nuclear translocation of resistance proteins, ROS-mediated degradation of alkaloid defense metabolites, and pathogen efflux pump overexpression that detoxifies phytoalexins. Compounding this vulnerability, epigenetic instability under drought alters H3K27me3 methylation patterns near resistance loci, potentially reactivating retrotransposons that disrupt *R* gene clusters, while carbon reallocation toward osmoprotectants depletes phenylpropanoid precursors essential for lignin-based fungal barriers. Consequently, engineered resistance deployed without climate-adaptive safeguards risks catastrophic failure in warming agro-ecosystems, necessitating the development of thermotolerant NLR analogs from extremophile plants (e.g., *Eutremasalsugineum*), redox-responsive promoters that activate defenses only during infection, and combinatorial stress phenotyping platforms simulating pathogen-pressure scenarios under projected climate regimes

10. CONCLUSION

Resistance for disease in plant can be created by use of genetically engineered tools that helps in providing pinpoint and selected treatment against infection. Despite its success, conventional breeding has long generation times and is susceptible to adaptive pathogens. However, the durability, specificity, and responsiveness of resistance genes have been significantly improved by CRISPR gene editing, gene stacking, and transgenesis. By using direct genome editing of plants, technologies such as CRISPR/Cas9 has proved to be extremely effective in promoting resistance against bacterial, fungal, and viral pathogens. Despite being highly effective, pathogen evolution, off-target behaviors, ecological risk, and consumer acceptability concerns pose challenges to genetic improvement for disease resistance. In order to release transgenic crops in a way that maximizes safety while maintaining consumer acceptability and environmentally sound agriculture, regulatory procedures and biosafety analysis remain essential. Future trends include integrated multi-omics, synthetic biology, and multigene stacking, which give crops long-term resistance and increase climate resilience.

REFERENCES

Abbas, M. S. T. (2018). Genetically engineered (modified) crops (Bacillus thuringiensis crops) and the world controversy on their safety. *Egyptian Journal of Biological Pest Control*, *28*(1), 1–12. DOI: 10.1186/s41938-018-0051-2

Agrios, G. N. (2005). *Plant pathology* (5th ed.). Academic Press.

Ali, Z., Mahfouz, M. M., & Piatek, A. (2022). CRISPR/Cas genome editing: Precision tools for improving plant disease resistance. *Current Opinion in Biotechnology*, *75*, 102710.

Buziashvili, A., Kvitko, B., & Dangl, J. L. (2023). Engineering synthetic immunity and integrating ecological strategies for sustainable crop protection. *Trends in Plant Science*, *28*(5), 432–445.

Chaloner, T. M., Gurr, S. J., & Bebber, D. P. (2021). Plant pathogen infection risk tracks global crop yields under climate change. *Nature Climate Change*, *11*(8), 710–715. DOI: 10.1038/s41558-021-01104-8

Cook, D. E., Mesarich, C. H., & Thomma, B. P. (2015). Understanding plant immunity as a surveillance system to detect invasion. *Annual Review of Phytopathology*, *53*(1), 541–563. DOI: 10.1146/annurev-phyto-080614-120114 PMID: 26047564

Dale, J., James, A., Paul, Y., Khanna, H., Smith, M., Peraza-Echeverria, S., Garcia-Bastidas, F., Kema, G., Waterhouse, P., Mengersen, K., & Harding, R. (2017). Transgenic Cavendish bananas with resistance to Fusarium wilt tropical race 4. *Nature Communications*, *8*(1), 1496. DOI: 10.1038/s41467-017-01670-6 PMID: 29133817

Dong, X., & Ronald, P. C. (2019). Genetic engineering for disease resistance in plants: Recent progress and future perspectives. *Plant Physiology*, *180*(1), 26–38. DOI: 10.1104/pp.18.01224 PMID: 30867331

Fuchs, M., & Gonsalves, D. (2007). Safety of virus-resistant transgenic plants two decades after their introduction: Lessons from realistic field risk assessment studies. *Annual Review of Phytopathology*, *45*(1), 173–202. DOI: 10.1146/annurev.phyto.45.062806.094434 PMID: 17408355

Gupta, S., Kumar, A., Singh, P. K., & Joshi, A. K. (2023). Leaf rust in cereals: Challenges and solutions. *Microbial Biotechnology*, *18*(4), 567–580.

Harding, R., Paul, Y., James, A., Smith, M., Kleidon, J., Shekhawat, U., Phillips, A., Kidanemariam, D., Dawit, A., & Dale, J. (2025). QCAV-4, the first genetically modified Cavendish (cv. Grand Nain) banana resistant to Fusarium wilt tropical race 4 approved for commercial production and consumption. *Plant Biotechnology Journal*, *23*(9), 3628–3637. Advance online publication. DOI: 10.1111/pbi.70178 PMID: 40488601

Haverkort, A. J., Boonekamp, P. M., Hutten, R., Jacobsen, E., Lotz, L. A. P., Kessel, G. J. T., Vossen, J. H., & Visser, R. G. F. (2016). Durable late blight resistance in potato through dynamic varieties obtained by cisgenesis: Scientifc and societal advances in the DuRPh project. *Potato Research*, *59*(1), 35–66. DOI: 10.1007/s11540-015-9312-6

Huang, K., Jahani, M., Gouzy, J., Legendre, A., Carrere, S., Lázaro-Guevara, J. M., González Segovia, E. G., Todesco, M., Mayjonade, B., Rodde, N., Cauet, S., Dufau, I., Staton, S. E., Pouilly, N., Boniface, M.-C., Tapy, C., Mangin, B., Duhnen, A., Gautier, V., & Rieseberg, L. H. (2023). The genomics of linkage drag in inbred lines of sunflower. *Proceedings of the National Academy of Sciences of the United States of America*, *120*(14), e2205783119. DOI: 10.1073/pnas.2205783119 PMID: 36972449

Islam, T., Watanabe, K., De Silva, N. I., & Hyde, K. D. (2024). Resistance mechanisms of plant pathogenic fungi to fungicides, environmental impacts, and sustainable solutions. [MDPI]. *Plants*, *13*(19), 2737. DOI: 10.3390/plants13192737 PMID: 39409607

Ji, W., Zhao, J., Liu, X., Deng, P., & Wang, C. (2025). Comparative transcriptomes reveal insights into different host responses associated with Fusarium head blight resistance in wheat. *BMC Plant Biology*, *25*(509). Retrieved from BMC Plant Biology.

Jones, J. D. G., & Dangl, J. L. (2006). The plant immune system. *Nature*, *444*(7117), 323–329. DOI: 10.1038/nature05286 PMID: 17108957

Klindworth, D. L.. (2023). Linkage drag analysis in Aegilops speltoidesintrogressions for wheat improvement. [Springer]. *Theoretical and Applied Genetics*, *136*, 168. DOI: 10.1007/s00122-023-04409-8 PMID: 37410182

Kumar, A. S. (2024). The ethical and security implications of genetic engineering. *Observer Research Foundation Issue Brief, 723*

Lee, S., Whitaker, V. M., & Hutton, S. F. (2016). Mini Review: Potential Applications of Non-host Resistance for Crop Improvement. *Frontiers in Plant Science*, *7*, 997. DOI: 10.3389/fpls.2016.00997 PMID: 27462329

Li, J., Li, Y., & Ma, L. (2021). Recent advances in CRISPR/Cas9 and applications for wheat functional genomics and breeding. *aBIOTECH*, *2*(4), 375–385. DOI: 10.1007/s42994-021-00042-5 PMID: 36304421

Li, T., Liu, B., Spalding, M. H., Weeks, D. P., & Yang, B. (2012). High-efficiency TALEN-based gene editing produces disease-resistant rice. *Nature Biotechnology*, *30*(5), 390–392. DOI: 10.1038/nbt.2199 PMID: 22565958

Ma, H., Liu, Y., Zhang, S., Sha, J., Sun, Y., Hu, Z., & Ma, H. (2025). Wheat resistance to Fusarium head blight and breeding strategies. *Crop Health*, *3*(1), 1–11. DOI: 10.1007/s44297-025-00048-1 PMID: 41649652

Mahanayak, B. (2024). Ensuring biosafety of genetically modified organisms (GMOs): Regulatory frameworks and risk assessments in India. *International Journal of Research Publication and Reviews*, *5*(7), 1368–1373.

Mahfouze, H. A., & El-Sayed, O. E. (2022). Comparative analysis of late blight resistance R genes and their coding proteins in some potato genotypes. *Czech Journal of Genetics and Plant Breeding*, *58*(1), 10–20. DOI: 10.17221/49/2021-CJGPB

National Research Council. (1993). *Managing Global Genetic Resources*. National Academies Press.

Noureen, A., Khan, M. Z., Amin, I., Zainab, T., & Mansoor, S. (2022). CRISPR/Cas9-mediated targeting of susceptibility factor eIF4E-enhanced resistance against Potato Virus Y. [Retrieved from Frontiers.]. *Frontiers in Genetics*, •••, 13.

Pan, S. (2024). *Bt crops – Definition, types, advantages, limitations. Biology Notes Online*. Retrieved from Biology Notes Online.

Ploetz, R. C. (2015). Fusarium wilt of banana. *Phytopathology*, *105*(12), 1512–1521. DOI: 10.1094/PHYTO-04-15-0101-RVW PMID: 26057187

Raj, M., Sharma, S., Gupta, A., & Nair, R. M. (2025). Mosaic Virus in Solanaceous Crops: Genetic Resistance and Biocontrol Approaches. *Tropical Plant Pathology*, *30*(2), 78–95.

Reddy, K., Patel, N., Singh, V. P., & Dasgupta, M. (2024). Root Knot Nematodes in Crops: Pathogen Dynamics and Mitigation Strategies. *Nematology Research Journal*, *12*(1), 45–60.

Rommens, C. M., Haring, M. A., Swords, K., Davies, H. V., & Belknap, W. R. (2007). The intragenic approach as a new extension to traditional plant breeding. *Trends in Plant Science*, *12*(9), 397–403. DOI: 10.1016/j.tplants.2007.08.001 PMID: 17692557

Sadhu, S., & Chatterjee, D. (2025). Advancing crop resilience: Plant breeding techniques to combat climate change challenges for small and marginal farmers. *Vigyan Varta*, *6*(3), 277–280.

Sánchez-Martín, J., Widrig, V., Herren, G., Wicker, T., Zbinden, H., Gronnier, J., Spörri, L., Praz, C. R., Heuberger, M., Kolodziej, M. C., Isaksson, J., Steuernagel, B., Karafiátová, M., Doležel, J., Zipfel, C., & Keller, B. (2021). Wheat Pm4 resistance to powdery mildew is controlled by alternative splice variants encoding chimeric proteins. *Nature Plants*, *9*(3), 502–503. DOI: 10.1038/s41477-023-01375-3 PMID: 36869130

Schaart, J. G., van de Wiel, C. C. M., Lotz, L. A. P., & Smulders, M. J. M. (2016). Opportunities for Products of New Plant Breeding Techniques. *Trends in Plant Science*, *21*(5), 438–449. DOI: 10.1016/j.tplants.2015.11.006 PMID: 26654659

Selvam, H., & Ragiman, S. (2024). *Eco-friendly Bt-based bioinsecticide for sustainable pest management*. Just Agriculture Newsletter.

Sharma, R.. (2024). Powdery mildew in plants: Epidemiology and control strategies. *Fungal Biology Reviews*, *58*(4), 312–328.

Sindhu, C., &Bala Janani, B. (2023). Empowering plant disease resistance through genetic engineering: A thorough analysis of current approaches, challenges, and future prospects. *International Journal of Scientific Development and Research*.

Suhag, D. (2024). *Regulatory and ethical considerations in biotechnology. Handbook of Biomaterials for Medical Applications* (Vol. 2). Springer.

Sun, K., Wolters, A.-M. A., Vossen, J. H., Rouwet, M. E., Loonen, A. E. H. M., Jacobsen, E., Visser, R. G. F., & Bai, Y. (2016). Silencing of six susceptibility genes results in potato late blight resistance. [Springer.]. *Transgenic Research*, *25*(5), 731–742. DOI: 10.1007/s11248-016-9964-2 PMID: 27233778

Trindade Maia, R., & Campos, M. A. (2021). Genetic variation: The source of biological diversity. *IntechOpen*, *19*, 1–15.

Wang, D., Mandal, P., Rahman, M. S., & Yang, L. (2025). Engineering tomato disease resistance by manipulating susceptibility genes. *Frontiers in Genome Editing*, *7*, 1537148. DOI: 10.3389/fgeed.2025.1537148 PMID: 39995605

Chapter 3
The Impact of Biodiversity on Plant Disease Dynamics in Organic Farming and Disease Management

Wasswa Shafik
http://orcid.org/0000-0002-9320-3186
Dig Connectivity Research Laboratory (DCRLab), Uganda

ABSTRACT

This chapter examines the role of biodiversity in shaping plant disease dynamics within organic farming systems and its implications for sustainable disease management. Biodiversity influences pathogen suppression, enhances soil microbial communities, and fosters ecological interactions that reduce disease incidence. It explores how diversified cropping systems, such as polycultures and crop rotations, contribute to disease resistance by disrupting pathogen life cycles and promoting beneficial microbial interactions. It also discusses the role of functional diversity in supporting natural enemies of plant pathogens, improving plant immunity through systemic resistance, and enhancing soil health. The chapter further evaluates how organic amendments, cover cropping, and habitat diversification affect plant-pathogen interactions and the effectiveness of biodiversity-driven disease management strategies. The discussion integrates current research, practical applications, and future perspectives on leveraging biodiversity to enhance resilience against plant diseases in organic farming.

DOI: 10.4018/979-8-3373-2429-6.ch003

1. INTRODUCTION

The use of biodiversity in organic farming contributes to pest and disease control. It is considered one of the essential functions of biodiversity. It is the management of biocontrols through the conservation of the natural enemies in diverse environments, which provides the most effective and long-lasting control measures. Such environments increase not only the range but also the number of natural enemies that can be host-specific. These tactics work well with small, generalist polytypic farms surrounded by a matrix of large intact ecosystems where obligatory biocontrols are found (Flachs et al., 2024). By contrast, complex ecosystems with a high rate of endemism, like those in New Zealand, where biocontrols are scarce, sustainable organic control using pest predators has a low chance of success. Research into the methods of enhancing local biodiversity in a mature native ecosystem has the potential to yield some, perhaps temporary, sustainable biocontrols. There are many important questions concerning the interaction of biodiversity and diseases that need to be addressed to develop effective management strategies in the natural world (Tiwari et al., 2024). The relationship between host group richness and reduced disease incidence was established well over a century ago, and this has many important implications for biodiversity and the provision of ecosystem services. This essay concentrates on some of the more recent and pressing issues that need addressing. Many plants have sufficient genetic diversity to be of importance in studies on the role of host diversity in the disease dynamics and resilience of ecosystems. Ecosystem disease theories are growing rapidly. We assume this huge expertise is available to the reader (Malik, 2024). We limit the discussion to the most widely studied ecosystem diseases, the fire blight of pear and other fruit trees, and the sudden oak death disease, which are caused by plant-pathogenic organisms, and sampled impacts of biodiversity as presented in Table 1.

Table 1. Impact of Biodiversity on Disease Dynamics in Organic Farming

Factor	Impact on Disease Dynamics	Benefits	Challenges	Application (s)	Reference
Crop Diversification	Reduces the spread of disease by varying host plants	Less susceptibility to pests and pathogens	Requires careful management, increased labor costs	Mixed cropping systems, intercropping in organic farms	(Wani et al., 2022)
Natural Predators & Pollinators	Predators control pest populations, reducing pathogen spread	Encourages natural pest control, enhances ecosystem balance	Need to ensure a stable population of predators	Ladybugs control aphids, and bees pollinate crops	(Adimas et al., 2024)
Soil Biodiversity	Enhances soil health, reduces pathogen build-up	Improves plant immunity, increases soil fertility	Soil management can be complex and slow to establish	Mycorrhizal fungi, soil microbes	(Phani et al., 2021)
Plant Variety & Disease Resistance	Increases genetic resistance to plant diseases	Promotes disease-resistant crops, reduces pesticide use	Limited access to diverse seeds, need for specialized knowledge	Heirloom varieties, disease-resistant crops	(Qin et al., 2024)
Agroecological Practices	Mimics natural ecosystems to maintain balance and minimize diseases	Reduces the need for chemical inputs, improves biodiversity	Requires larger land areas, requires knowledge of ecosystem balance	Rotational grazing, cover crops for soil health	(Devkota et al., 2024)
Resilient Crop Systems	Diverse systems provide ecological buffers against disease	Reduces dependence on synthetic chemicals, increases resilience	Initial transition costs, adjusting to local environments	Organic farms with crop rotation, polyculture	(Lovell-Read et al., 2023)
Soil Health	A strong soil microbiome prevents pathogenic growth	Promotes healthy plant growth, reduces disease spread	Requires regular soil monitoring, time-consuming	Composting, reduced tillage practices	(Alae-Carew et al., 2022)
Climate Resilience	Biodiversity provides ecosystems that can adapt to changing climates	Improved plant resistance to climate stress, disease control	Changes in climate can alter disease patterns unpredictably	Drought-resistant crops, temperature-tolerant varieties	(Liu et al., 2021)

continued on following page

Table 1. Continued

Factor	Impact on Disease Dynamics	Benefits	Challenges	Application (s)	Reference
Integrated Pest Management (IPM)	Uses a combination of biological, mechanical, and cultural practices to control pests	Minimizes chemical use, enhances natural pest control	Requires constant monitoring, knowledge of pests	Crop rotation, use of biological pesticides like neem oil	(Carney, 2021)
Ecological Balance	Maintaining biodiversity in farming systems helps regulate disease outbreaks	Natural pest control reduces the spread of pathogens	Balancing all components of the ecosystem	Integrated farms using wildlife corridors, wetlands	(Chiti et al., 2024)

2. UNDERSTANDING BIODIVERSITY IN AGROECOSYSTEMS

Biodiversity, the variety of life in all forms, encompasses a number of levels of organization at which it unfolds. It occurs at the ecosystem level, where the range of living organisms observes it in terms of community composition and function, and at the genome level, where the genetic variability among genotype populations gives an idea of the variety of traits that may influence the growth, development, and production of these populations. Different components are found at intermediate levels of species diversity, which are generally categorized taxonomically into various groups that belong to the common life forms that form an ecosystem (Fontaine et al., 2024). Genetic diversity refers to the range of different genes carried by a species, including the DNA and other inherited cellular constituents. Different forms of biodiversity are correlated, and a change in one type of biodiversity can result in a change in others. In the natural world, diversity can combine with numerous species and interactions to achieve the provision of various ecosystem services, including natural control over diseases. Firstly, genetic diversity within a species can help to reduce the severity of some specific diseases; for example, increased host plant diversity leads to reduced severity of diseases. Second, the number of species as a component of diversity that can potentially influence ecosystem functioning is positively correlated with ecosystem stability and productivity (Guldenring et al., 2024). Third, increasing spatial heterogeneity in ecosystem conditions, increasing horizontal and vertical components of inter- and intra-species diversity, changes in the niche of each species, and limiting their transmission reduce the rapid build-up and spread of diseases, as well as the maintenance of genetic integrity that results from genetic mobility, using artificial intelligence, as presented in Figure 1.

Figure 1. Artificial intelligence (AI) benefits in Agriculture

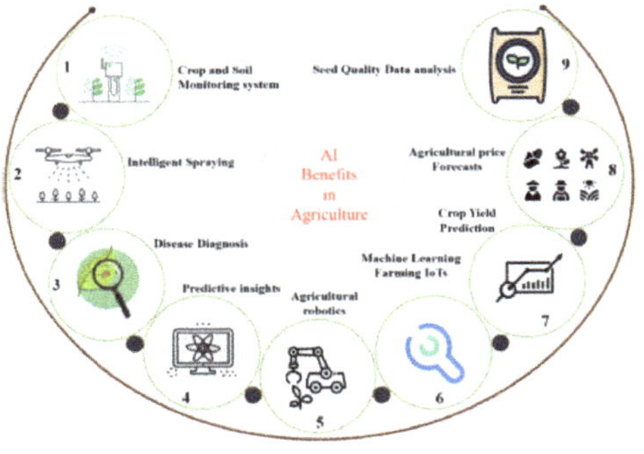

2.1. Definition and Components of Biodiversity

Biodiversity is an intrinsically complex and multifaceted concept. In the agricultural context, it is about much more than the inventory of species within a field; it also includes the differences among them and the complex interactions that occur. Biodiversity in an agroecological context is composed of three main components:

(i) genetic diversity at the intra- and inter-specific scales within and among agricultural fields;
(ii) species diversity associated with the number of plant, crop, non-crop, and animal species living on and around the agricultural fields; and
(iii) the multitude of ecological processes that emerge from the relationships between these species and their supporting ecosystems.

Genetic diversity contributes to the assimilation of nutrients, to the development of positive relationships between different biological species, and to boosting the entire system. Species diversity, also called functional diversity, has a significant influence on ecosystem properties, driving ecosystem functions and services that contribute to agroecosystem health and functioning. Ecosystem services, or the benefits people obtain from ecosystems, include directly the goods and services such as food production, pest regulation, pollination, as well as non-material services like aesthetics or recreational value. Many relationships exist among biodiversity components (Ismail et al., 2023). Among these interactions, genetic diversity within a species can enhance species richness and ecosystem functioning. Biodiversity contributes to increasing the stability of the functioning of living systems and their resilience as a capacity to bounce back after a disturbance. It also helps maintain the capacity of living systems to adapt to new situations. Biodiversity contributes to making agroecosystems more resilient to disturbances, particularly for plant disease attacks. Inversely, changes in biodiversity, particularly losses, can lead to increasing vulnerabilities to plant disease outbreaks (Shafik, 2024a).

2.2. The Role of Biodiversity in Disease Suppression

Rapid breakouts of plant disease often occur in agroecosystems where extensive monocultures have been simplified. An ecosystem can be envisioned as an interacting complex of organisms interconnected in chains of energetic or material flows or networks. The diversity of organisms in such a system leads to increased stability. Balanced ecosystems are consequently conducive to disease suppression by controlling pathogens. Leguminous plants, for example, are known to create rhizosphere communities abundant in root pathogens and plant growth-promoting rhizobacteria. Complex plant communities can reduce the disease-suppressive activities of the bulk and rhizosphere soils. The diversity can stimulate soil microbial communities, boosting suppressiveness against pathogen attack (Shafik, 2025b). The ability of a diverse plant community to limit disease outbreaks is conceptualized as the competitive exclusion of the soil pathogens by microbial antagonism. Competition among organisms in an environment can limit the outbreak of disease. Higher microbial diversity in a soil has also been shown to suppress soil-borne diseases, both abiotically and biotically. Organisms occupying higher trophic levels above the antagonists of the pests also impact disease. Predators and parasites of pests can naturally suppress their populations (Kovács et al., 2024; Shafik, 2024b). The antagonists occupy links within a food web that potentially allow the bottom-up impacts of biodiversity to influence disease indirectly. Field-scale studies have convincingly demonstrated the benefits of intraspecific crop genetic diversity in reducing the severity of disease. There are many successful strategies for exploiting biodiversity in organic farming, many of which involve habitat manipulations (Fahad et al., 2021; Shafik & Kalinaki, 2024). Integrating beneficial plants directly into the cropping system is an especially enticing prospect for organic farming, as sampled organic practices and disease management in Table 2.

Table 2. Organic Farming Practices and Disease Management

Organic Practice	Impact on Disease Management	Benefits	Challenges	Examples	Reference
Crop Rotation	Prevents the build-up of soil-borne diseases by alternating crops	Reduces pathogen populations, improves soil health	Requires knowledge of compatible crops and rotation schedules	Rotating between legumes and cereals in organic farms	(Shafik, 2025c)
Composting	Enhances soil health and reduces plant diseases by promoting beneficial microorganisms	Improves soil fertility, reduces reliance on chemical fertilizers	Time-consuming, requires proper management of compost	Composting plant residues, animal manure for soil health	(López-García & Carrascosa-García, 2023)
Cover Cropping	Provides ground cover to prevent soil erosion and suppress weeds	Improves soil health, prevents weed-borne diseases	Requires careful selection of cover crops, management effort	Clover or rye as cover crops to suppress weeds	(Orounladji et al., 2024)
Biological Pest Control	Uses natural predators, pathogens, or parasites to manage pests	Reduces chemical pesticide use, supports biodiversity	Limited availability of natural predators, higher initial cost	Release of beneficial nematodes, ladybugs for pest control	(Shafik, Tufail, Liyanage De Silva, et al., 2025)
Use of Resistant Varieties	Incorporating disease-resistant crops reduces pathogen impact	Minimizes pesticide use, ensures healthy crop production	Limited availability of resistant varieties for all crops	Disease-resistant tomatoes, wheat, and other crops	(Shafik, Tufail, Liyanage, et al., 2023)
Mulching	Helps suppress weeds and retain soil moisture, reducing plant stress	Enhances soil health, limits pathogen spread	Requires regular replenishing, can attract some pests	Straw, wood chips, and organic mulches for weed control	(Flachs et al., 2024)
Reduced Tillage	Minimizes soil disruption and damage to beneficial organisms	Prevents erosion, enhances soil structure, and health	Requires more advanced equipment and expertise	No-till farming, direct seeding	(Tiwari et al., 2024)
Green Manuring	Incorporates plant material into the soil to improve soil health and suppress weeds	Enhances soil fertility, increases biodiversity	Requires labor and timing for planting and incorporation	Using legumes like clover or vetch for soil nitrogen	(Malik, 2024)
Water Management	Reduces waterlogging and promotes healthy plant growth	Prevents waterborne diseases, maintains crop health	Requires careful irrigation management	Drip irrigation, rainwater harvesting	(Shafik, 2025a)

continued on following page

Table 2. Continued

Organic Practice	Impact on Disease Management	Benefits	Challenges	Examples	Reference
Companion Planting	Pairing plants to repel pests and reduce disease spread naturally	Enhances plant health, reduces chemical use	Requires understanding of plant compatibility	Basil and tomatoes for pest control, marigolds for nematodes	(Tiwari et al., 2024)
Organic Pesticides	Uses non-synthetic substances to control pest and disease outbreaks	Reduces environmental impact, safer for humans and wildlife	May be less effective than synthetic pesticides, costlier	Neem oil, garlic, or chili sprays, diatomaceous earth	(Shafik, Tufail, Namoun, et al., 2023)
Soil Fertility Management	Ensures the soil has the necessary nutrients to support strong plant health	Strengthens plant resistance to diseases	Requires constant monitoring and adaptation to soil needs	Application of organic fertilizers like bone meal, fish emulsion	(Phani et al., 2021)

3. PLANT DISEASE DYNAMICS IN ORGANIC FARMING

Plant diseases are known to develop fast in organic farming systems. Disease dynamics in organic systems take place in the complex interactions of various biotic and abiotic factors. Organic farming systems depend on the ecological approach as well as on social and economic sustainability, and therefore, they do not rely on the application of synthetic chemicals. This leads to a number of limitations on conventional plant disease management (Lovell-Read et al., 2023). In organic farming, agricultural practices cannot interfere with the functioning of general ecological properties, such as climate, soil health, and biodiversity. Furthermore, organic farming emphasizes the importance of the interactions among biotic organisms. The continuous dynamics of co-evolutionary interactions among plants and pathogens are based on responses to both plant and pathogen-associated genotype-by-genotype interactions (Seipel et al., 2019). These interactions are influenced by environmental factors that condition plant-inherent metabolomic profiles. In organic farming systems, the interactions above are further complicated by the enhanced natural selection resulting from the setup of relatively uniform environments. However, the overall mechanisms of the plant-pathogen dynamic are poorly characterized to date. This complex pattern is further complicated by considering the actual mechanisms of the plant-pathogen dynamics in a farming system where organic practices maintain the conservation of genetic variability by relying on open planting materials and a range of adapted crop species and landraces (Carney, 2021). Even though some closing of this knowledge gap has occurred in recent years, there is little research that defines plant disease dynamics considering the overall environmental pressure and genetic diversity of organisms involved. Therefore, in this review, plant disease dynamics in relation to the seed system and adaptation to organic farming contexts are discussed. Concerning the literature review, relevant studies were selected through a search on a database (Alae-Carew et al., 2022). We carried out separate searches that included combinations of the following keywords: dynamic or dynamics, sustainable, ecology, and farming system.

3.1. Challenges of Disease Management in Organic Systems

Disease management is an important aspect of plant production in organic systems where the use of synthetic pesticides is often legally restricted. Organic farmers must rely on disease management strategies such as sanitation and habitat manipulation to control plant pathogens and pest populations. Maintaining plant health is, however, a complex task given that organic production standards restrict the use of chemical fertilizers and plant protection products, and consequently, nutrient availability to plants is limited (Shafik et al., 2024). There is some evidence indicating that plants grown under organic conditions may indeed be more susceptible to pathogen infection, although breach of the plant cuticle is considered the key event facilitating pathogen establishment in planta. In addition, for fields under organic management, environmental conditions might also create disease-conducive scenarios, including the presence of particular weed species and their associated pathogen communities at the edges of fields, or geographic isolation allowing home-field disease escalation through pathogen adaptation (Shafik, Tufail, De Silva, et al., 2025). In certain instances, plant disease, driven by environmental conditions, has profound negative social and economic consequences for our societies, a fact that is particularly true in the case of the organic sector, which supports its farming systems using economic and political tools, as some factors are presented in Figure 2.

Figure 2. Agricultural Sustainability

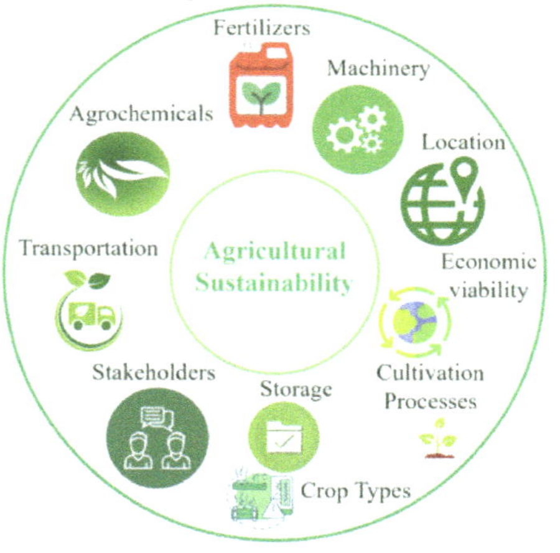

For the organic sector, it is of great importance to demonstrate that efficient and effective disease management options are available, and it is here that research and education become important. Now more than ever, an integrated approach to disease management is advocated for in order to meet the challenges of plant disease in the twenty-first century (Shafik & Azrour, 2024). This approach takes advantage of different disease control options already available as well as new research currently underway that promotes healthy plants with a high level of natural resistance to disease, thereby reducing the chance of outbreaks occurring. Ultimately, plant disease-resistant varieties, i.e., cultivars that are partially or fully protected against one or more plant pathogens, are usually employed in the field on their own, although a more successful approach could be the evolution of disease-resistant varieties designed for use in a broader crop protection strategy. In effect, they would not use an either/or philosophy, but would manage crop protection using numerous technologies that work in a complementary manner, for example, combining the deployment of varieties with resistance against two or more pests in an attempt to maintain product differentiation in the marketplace (Singh et al., 2024).

3.2. Factors Influencing Disease Dynamics in Organic Farming

Disease dynamics in organic farming are determined by a complex array of factors, some of which are unique to organic systems. Soil health, rich in organic matter, supports plant nutrition, and root stress may lessen the impact of some diseases, moderating disease patterns. Soil health is also a strong determinant of the composition and function of the soil microbiome and virome. This living community is shaped by conditions, including soil type and climate, including weather conditions at the time when the community was established. Crop diversity or monoculture can alter disease dynamics in systems where diseases are primarily soilborne (Lovell-Read et al., 2023). The climatic conditions experienced by crops over the season can, in some cases, greatly influence disease outcomes. Interactions also occur between pests and pathogens in response to environmental changes. These particularly organic-farming-specific drivers are classed as exogenous because they stem from the characteristics of the system rather than management practices. With the major influence on organic farming disease dynamics of properties of the system itself, being informed about all prospective relevant drivers of change can provide key insights into seasonal disease dynamics that can help growers manage the risks that they face (Seipel et al., 2019). The role of some of these drivers can be imperative, such as the link between soil temperature and clubroot and the influence of seasonal conditions on disease severity in rotations versus continuous crops. Counterbalancing these drivers are management practices that can either exacerbate patterns, such as the continuous or repeated establishment of host plants or the suppression of local diversity in pests, diseases, and weeds, or on the contrary contain or suppress these drivers, such as using a strategic crop break to allow the damage pool to dissipate, cover cropping, and using compost deficiencies to select out soilborne diseases (Fontaine et al., 2024). Here we will examine some of the key variables driving disease dynamics in organic farming in more detail.

4. INTERACTIONS BETWEEN BIODIVERSITY AND PLANT DISEASES

The relationship between biodiversity and plant diseases has been the focus of much research. Evidence suggests that there is a mutual dependence linking diversity and disease dynamics in such a way that the latter can alter the former, and vice versa. Both negative and positive feedback loops operate in this relationship: the presence of biodiversity may decrease disease pressure through the action of natural enemies or through other interference interactions, while increased diversity may provide more niches for the presence of additional pathogens or aphids, thus increasing disease incidence (Flachs et al., 2024). For instance, it has been shown that in fields of mixed cabbages, one species may act as a disease trap crop. This has an indirect positive effect: protecting the other species of cabbage against the disease by its present epidemiological structure. It stands to reason, therefore, that increased connectivity between epidemiological units can have a negative influence. Also, in the case of conserved tillage practices, the suppression of aerogenic dispersal has a beneficial effect because the disease is primarily spread by wind. Biodiversity can introduce many unwanted pests and diseases: for instance, pesticide use resulted in the selection of detrimental species and vectors on organic farms that were under a regime of low pesticide input (Wani et al., 2022). In turn, increased overwinter survival due to pesticide use acting as an 'ecosystem service' was responsible for a switch from summer to winter disease. There are, therefore, many different ecological mechanisms by which human interventions can be related, oftentimes indirectly, to changes in disease status. The consequences of these joint interactions are ecosystem processes and ecological patterns that, in turn, mediate biological control. Such processes range from enrichment of soil nutrient content to attract fungivorous fauna, through to effects of landscape composition and configuration on the distribution of plant habitats, parasites, diseases, herbivores, predation, and herbivory (Qin et al., 2024). Clearly, there are many pathways by which experimental and theoretical studies on disease dynamics can contribute to an understanding of disease management strategies in agroecosystems.

4.1. Positive and Negative Feedback Loops

Positive, negative, and double feedback loops make the relationship between biodiversity and plant diseases intricate. Positive feedback loops emphasize that increased biodiversity can lead to better ecosystem functioning, which can result in more efficiently released protective services and disease suppression functions, resulting in lower intensity and scale of plant disease epidemics. In the context of pest management, for example, studies show that high levels of biodiversity improve ecosystem functions such as natural pest control and increase the beneficial services of restoring plant resistance through biologically induced systemic resistance in plants (Liu et al., 2021). Thus, increased biodiversity can also be envisaged as expanding the plant phyllosphere diversity, which can then manage the plant diseases more efficiently, for instance, via the homogenized assembly of beneficial ECM fungi, which, besides allowing pathogen suppression, may either lead to reduced biocontrol fungi via priority effects, commensalism, or competition. By exploring the links between biodiversity and taxonomic and trait diversity, our analysis goes beyond the bottom line of disease reduction. While increasing the proportion of natural habitat in the landscape might enhance biodiversity-disease relations, the included publications qualifying negative and double feedbacks can argue for both the effectiveness and ecological risks of maintaining or changing high biodiversity profiles. We argue here for a more holistic study of biodiversity, ensuring the analysis of positive, negative, as well as double feedbacks, instead of simply emergency responses (Carney, 2021). Based on the present review of feedback loops, we offer some key topics that should be considered when integrating biodiversity and disease management, in our effort to transform insight into action and practice.

5. CASE STUDIES AND RESEARCH FINDINGS

5.1. Case Study 1: Organic Farming

United Kingdom, England: In an extensive survey of organic farms, diversification of crop and non-crop species was found to be the major factor influencing levels of pests and diseases, even including the use of manure. In a specifically designed experiment, cereal species and varieties were tested for effects on levels of Barley Yellow Dwarf Virus. Soil properties, moderate tillage, and the availability of mineral nitrogen were also found to be important in determining levels of the virus. Italy, Palermo, Sicily: In two separate studies over three and one year(s), respectively, the spatial distribution of different wheat blast symptoms was related to their proximity to other inter-planted crop species and neighboring forest (Chiti et al., 2024). The continuous nature of the land use structure may have accounted for the perceived lack of significant effects. Austria: In two experiments—one carried out in 1854 and the second in 2007—plots of White Lady potatoes were surrounded by different types of plant species. The level of late blight was assessed from a combination of symptoms and defoliation. The German version of the introduction does not include methodologies; therefore, it is not possible to determine the overall results or different factors. Netherlands: In an analysis of the impact of bumblebee species on the presence of the pathogen, the collection of pollen from field-grown crops was noted to contain some bees with sporulation of the pathogen (Guldenring et al., 2024). However, when comparing a reference to the mean values, no significant effects were observed.

5.2. Field Experiments on Biodiversity and Disease Dynamics

Evidence that maintaining or enhancing biodiversity can effectively reduce disease is rapidly accumulating from field trials. Such trials are generally designed experimentally along a gradient of species levels or combinations of treatments, and documentation occurs over a number of years. The measures that are taken to characterize disease losses usually represent key aspects of disease: prevalence, or the proportion of diseased individuals; and severity, or the extent of symptoms in infected individuals. The field experiments were developed from theory and established ecological practice, with mixtures of perennial grasses for lowland pastures and organically managed cereals (Flachs et al., 2024). Disease has been shown to be prevented or to increase with sown diversity. Both of these responses can be associated with particular ecological principles and, therefore, are valuable underpinnings of these principles. In semi-natural grassland, disease losses can be substantial. From the absence of disease, we can see a reduction in disease severity in plots with the most diverse components. In more recently sown grass mixtures for cereals, we have shown that we can boost disease losses in the most experimentally manipulated components, so that in pure stands, losses are twice as great as those in mixed stands. The protection was shown to be due to a reduction in pathogen movement from diseased individuals to new host individuals, effectively isolating infected samples (Tiwari et al., 2024). These are tiny-scale local ecological processes, and the demonstration of enhanced protection in the plots illustrates the practicality of using this newly developed knowledge to guide growers. The accompanying significant commercial backing is a demonstration of the potential benefits of identifying and subsequently manipulating these ecological blocks in disease epidemiology. This research establishes the power of diversity to prevent pathogen transmission. Even though we are unable to use precise manipulations to increase disease, the control that is designed based on an integrated model is able to show rapid rates of disease spread with no disease present in surrounding landscapes (Malik, 2024).

6. IMPLICATIONS FOR ORGANIC FARMING PRACTICES

Dividing plant disease dynamics from natural systems and community-level populations has allowed us to probe links between ecology and disease, but knowing how to apply this knowledge to the agenda of organic farming is far from trivial. This naturally raises the question of where to turn in this final paragraph of an attempt to connect theory and application. Certainly, relevant to research design are the practical considerations of farmers and conservation managers. We argue that biodiversity-oriented strategies can and should be included as options for these relevant stakeholders. From a sustainable agriculture perspective, the most obvious reason for doing so is that they will add resilience (Adimas et al., 2024). But at a more pragmatic level, biodiversity management is integral to these strategies. Farmers or stakeholders uninterested in biodiversity for its own sake are likely to incorporate such practices if there are clear benefits for them and standard organic agriculture performance. In essence, this means grappling with the bundle of questions that ultimately hinge on farmer adoption. What do stakeholders need to incorporate into their management toolbox? Some ingredients that enhance diversity. Various management practices can serve to increase and protect conventional biodiversity in farming systems. For example, landscape and tillage diversification, the inclusion of cash and non-cash crops in rotation, as well as within-field diversification, can accentuate the positive effect of farming system diversity on biodiversity at each scale (Phani et al., 2021). Intercropping systems can also increase species diversity when species richness is measured at the scale of crop plots. But diversity itself does not necessarily promote species richness among agro-ecosystem types at the community or all levels. Intercropping of barley and canola/wheat combinations tended to enhance abundance and species richness of invertebrates over that in the sole cropping systems. Such enhancements of diversity relative to the appropriate comparator for that scale are likely to affect natural enemy populations and their efficacy. Biodiverse systems and living mulches can promote natural enemy species diversity and richness, and sometimes abundance, in field crops in the spring. The use of native species and plants with an attractive appearance for the diversification of habitats and hedgerows can also be important (Devkota et al., 2024). Those who know what is appealing to their neighbors can contribute to the enhancement of diversity by engaging in educational activities focused on others.

6.1. Incorporating Biodiversity Conservation Strategies

The incorporation of agroecological and conservation concepts on farms may come in different forms that are adaptable to any agricultural context, with the guideline "Sow diversity to sow resistance." As generic solutions are of little help, some operational steps and considerations could be shared.

(1) Creating habitats and favoring biodiversity may often materialize as the addition of native margin strips. Many farmers have abandoned on-farm herbaceous elements or even entire old meadows or wetlands; by reintroducing native herbaceous species, including functional perennials and woody elements, these may be reestablished, too (Liu et al., 2021). In arable fields, summer crops such as fallow and oilseed hemp can also provide pollinator-friendly habitats; winter flowers can be planted in potato fields and in fallow winter arable fields.
(2) Create buffer strips and native woody plants near the farm's fence as an additional habitat for disease-suppressing organisms.
(3) Stop using plant protection products once every 3–5 years in order to reduce the selection pressure on pathogens and pest biota, thus improving the sustainability of the conventionally exploited environment. In summary, relevant biodiversity cannot be established as a one-action event, but must be an integral farm objective.

When designing on-farm biodiversity actions, farmers should aim to

(i) maintain diversity for multiple functions,
(ii) use a multi-scale implementation of biodiversity conservation, and
(iii) promote a synergy of solutions between increased biodiversity and improved management (such as highways for beneficials and flower strips between the crop species).

By selecting for a diversity of species, functional biodiversity will be enhanced at the same time. Agroecologists argue that ecosystem health emerges from the increasing diversity of biodiverse adaptive ecosystem responses, and there is no specific biodiversity-health relationship. Instead, causation might go in the opposite direction; a biodiverse adaptive ecosystem response enhances crop health. A growing ecological insight is that the addition of a single temporal or spatial component does not necessarily create a healthier or more resilient farming system (Flachs et al., 2024). Co-integrative biodiversity conservation opens the way whereby farmers cannot only foster food production but also build healthy, resilient farming systems without needing to worry about conserved species for future generations. The examples below show how biodiversity conservation practices can also be useful disease management strategies to integrate into organic systems, mentioning benefits for human health, crops (creating habitat for pest and disease biocontrol), or rangeland (enhancing resistant sheep populations) (Wani et al., 2022). Relying on wild species has been cited as a more transgressive—but promising—approach, exposing agroecologically managed farmers to the economic risk of having no backup in case of external pressures leading to biodiversity loss. Although there will always be constraints on land availability, we conclude that incorporating an on-farm biological value also into the non-cropped part of the farm is the backbone of biodiversity promotion in organic farming. We see the first step of organic farming as health-promoting soil management, followed by multi-purpose on-farm biodiversity conservation (Qin et al., 2024).

7. FUTURE DIRECTIONS AND RESEARCH NEEDS

Clearly, there has been much progress in our understanding of how biodiversity influences disease and contributes to disease management in organic farming. End-to-end interdisciplinary approaches that aim to understand functioning at different levels from ecology to agronomy and from farmer practices to social relationships have the potential to enhance our knowledge further. Current management strategies should be encouraged and applied, but autonomous mechanisms that reinforce these practices, for which biodiversity is crucial, need to be identified and understood.

7.1. Current management strategies

In this regard, long-term studies that allow for the incorporation of biodiversity conservation strategies into farming systems need to be encouraged in order to assess their long-term impact on disease dynamics and production. Additional support for the identification and testing of innovative farming practices, technologies, and methods aimed at multifunctional sustainability can help support biodiversity conservation in agricultural systems and diversification with agroecological infrastructure. However, there are still important areas with poor knowledge that need to be addressed. For example, we know little about the implications of not conserving artificial landscapes or the effects of urban gardens or greenhouses (Lovell-Read et al., 2023). It is still unclear which type of agrobiodiversity should be restored in order to enhance biodiversity in food production systems. Needs for further science are therefore identified, as exemplified by the following five questions:

(1) What is the relationship between landscape variation, disease or pests, beneficial organisms, healthy food systems, and income?
(2) What is the relationship between farm-scale variation, disease or pests, beneficial organisms, healthy production, and income?
(3) Can we model this so that the important relationships are quantifiable, scalable, and generalizable in time and space, and can improve farmers' income?
(4) Can we develop autonomies that will help us to adapt our models over time to control pests and diseases more effectively in organic farming and food systems?
(5) Can we show this to farmers, non-governmental organizations working on food and agriculture, and policymakers to ensure the development of better socio-ecological systems that address biodiversity at the farm and landscape scale?

In doing so, we can develop agroecological ecosystems that are resilient to pest and disease emergence, thereby contributing to a safe, healthy, and sustainable global food system.

7.2. Advancing Understanding of Biodiversity-Disease Interactions

Elucidating the dynamics of the complex biodiversity-disease interactions is paramount for advancing our thinking on the ecological origins of the biodiversity-disease relationship. To achieve this, we need to carefully design research studies to disentangle biodiversity-disease relationships that operate at different ecological scales. Such basic information is key to understanding the constraints put on new biodiversity management strategies, especially in organic crops (Liu et al., 2021). Using the latest and developing technologies, spanning from rapid DNA-based crop disease sensors for use in the field to state-of-the-art genomics and data analytics, we need to unravel the complexity of multiple environmental factors and complementary multidisciplinary approaches that feed into the overall research program. This interdisciplinary work should be conducted in consultation with students and practitioners to ensure that it is embedded in the ecosystem, to ensure improved knowledge and good dissemination of the research findings to farmers and advisors (Qin et al., 2024). To foster exchange between research and the grassroots level of horticulture, examples of innovative disease management strategies should be integrated into applied research contracts. Agriculture is only going to make advances in overall sustainability if we combine practical on-farm experiences with scientific research and new technologies. With biodiversity next, we can start to improve biodiversity-disease management. We need to understand better the biodiversity-disease relationship, spanning the normally studied plant diversity to include in-field predators, parasitoids, cover crops, and intercropping at the plot level (Tiwari et al., 2024; Wani et al., 2022). We would look at how such increases in predator and parasitoid assemblages would increase the ability of predators and parasitoids to suppress key pest populations in order to provide additional control in organic or agroecological systems.

8. CONCLUSION

Although most of the research has been conducted in model systems, together with what has generally been proven in pathology, further contributions from neighboring disciplines such as entomology have synthesized general strict knowledge to indicate that there is a very high potential in promoting resilience and sustainability by implementing systems with high levels of biodiversity in general and functional biodiversity in particular. The importance of looking at the different spatial and temporal scales in which different types of biodiversity can have different types of impact was also highlighted. Moreover, although the present-day state of the art indicates that the dynamics of populations in interactions such as these are highly complex and thus provide both considerable challenges and opportunities for smart and knowledge-based disease management, the complexity alone is no excuse for more knowledge and better handling of systems in promoting this knowledge. What has been shown so far clearly points to the many possibilities in this domain. Despite many findings and insights opening doors to new ideas and possibilities for promoting healthy crops based on ecological principles, it is also clear that more research and demonstrations are required to apply such knowledge in organic and other kinds of farming. As biodiversity is highly valued as a goal to be attained for conservation of the wider environment, conservationists seem to mainly strive for enhancing existing systems or promoting the establishment of new habitats used by a wide array of species. Furthermore, systems that allow for a high level of breeding, such as is required for the production of many kinds of grass seeds, booming after genetic modifications and others, do not seem to contain the right qualities for disease-suppressive interactions. Also, agricultural systems aiming at increased biodiversity in this respect will have to integrate biodiversity conservation at least continually to optimize the health-promoting biology in the systems based on ecological interactions. Hence, we suggest that stakeholders of this attractive and fascinating domain match their forces and collaborate on this issue. We need to try to integrate farmers, researchers, food companies, and consumers to establish a better understanding of the value of high diversity systems in a spatial and temporal frame of reference, and also in obtaining food security, as to what the role of these systems is going to be. In other words, let's keep it on the agenda!

REFERENCES

Adimas, Z. T., Adimas, M. A., & Abera, B. D. (2024). Plant-based bioactive compounds for grain storage: a comprehensive review. In *Cogent Food and Agriculture* (Vol. 10, Issue 1). DOI: 10.1080/23311932.2024.2316152

Alae-Carew, C., Green, R., Stewart, C., Cook, B., Dangour, A. D., & Scheelbeek, P. F. D. (2022). The role of plant-based alternative foods in sustainable and healthy food systems: Consumption trends in the UK. *The Science of the Total Environment*, *807*, 151041. Advance online publication. DOI: 10.1016/j.scitotenv.2021.151041 PMID: 34673070

Carney, J. A. (2021). Subsistence in the Plantationocene: Dooryard gardens, agrobiodiversity, and the subaltern economies of slavery. *The Journal of Peasant Studies*, *48*(5), 1075–1099. Advance online publication. DOI: 10.1080/03066150.2020.1725488

Chiti, T., Blasi, E., & Chiriacò, M. V. (2024). Carbon sequestration in a bamboo plantation: A case study in a Mediterranean area. *Journal of Forestry Research*, *35*(1), 51. Advance online publication. DOI: 10.1007/s11676-024-01696-9

Devkota, K., Ferreira, A. B., Timberlake, T. P., & dos Santos, C. F. (2024). The impact of pollinator decline on global protein production: Implications for livestock and plant-based products. *Global Ecology and Conservation*, *50*, e02815. Advance online publication. DOI: 10.1016/j.gecco.2024.e02815

Fahad, S., Saud, S., Akhter, A., Bajwa, A. A., Hassan, S., Battaglia, M., Adnan, M., Wahid, F., Datta, R., Babur, E., Danish, S., Zarei, T., & Irshad, I. (2021). Bio-based integrated pest management in rice: An agro-ecosystems friendly approach for agricultural sustainability. In *Journal of the Saudi Society of Agricultural Sciences* (Vol. 20, Issue 2). DOI: 10.1016/j.jssas.2020.12.004

Flachs, A., Bastos, C., Heath, D., & Venkateswar, S. (2024). Introduction to Special Collection: Plant-Anthropo-Genesis: The Co-Production of Plant–People Lifeworlds. *Journal of Ethnobiology*, *44*(1), 3–10. Advance online publication. DOI: 10.1177/02780771241228068

Fontaine, S., Abbadie, L., Aubert, M., Barot, S., Bloor, J. M. G., Derrien, D., Duchene, O., Gross, N., Henneron, L., Le Roux, X., Loeuille, N., Michel, J., Recous, S., Wipf, D., & Alvarez, G. (2024). Plant–soil synchrony in nutrient cycles: Learning from ecosystems to design sustainable agrosystems. In *Global Change Biology* (Vol. 30, Issue 1). DOI: 10.1111/gcb.17034

Guldenring, R., Andersen, R. E., & Nalpantidis, L. (2024). Zoom in on the Plant: Fine-Grained Analysis of Leaf, Stem, and Vein Instances. *IEEE Robotics and Automation Letters*, *9*(2), 1588–1595. Advance online publication. DOI: 10.1109/LRA.2023.3346807

Ismail, B. I., Sehmi, M. N. M., Ahmad, H., Baharom, S. H., & Khalid, M. F. (2023). Robotic Research Platform for Agricultural Environment: Unmanned Ground Vehicle for Oil Palm Plantation. *Journal of Cases on Information Technology*, *25*(1), 1–32. Advance online publication. DOI: 10.4018/JCIT.328579

Kovács, T., Botos, S., & Felföldi, J. (2024). What does digital transformation do in agriculture? A systematic literature review of Agri-food sector's digitalisation. *Agrárinformatika Folyóirat*, *14*(2). Advance online publication. DOI: 10.17700/jai.2023.14.2.707

Liu, L., Wang, J., Wang, F., & Yang, X. (2021). The impact of the planting of forest biomass energy plants under the embedded Internet of Things technology on the biodiversity of the local environmental ecology. *Environmental Technology & Innovation*, *24*, 101894. Advance online publication. DOI: 10.1016/j.eti.2021.101894

López-García, D., & Carrascosa-García, M. (2023). Agroecology-oriented farmers' groups. A missing level in the construction of agroecology-based local agri-food systems? *Agroecology and Sustainable Food Systems*, *47*(7), 996–1022. Advance online publication. DOI: 10.1080/21683565.2023.2217095

Lovell-Read, F. A., Parnell, S., Cunniffe, N. J., & Thompson, R. N. (2023). Using 'sentinel' plants to improve early detection of invasive plant pathogens. *PLoS Computational Biology*, *19*(2), e1010884. Advance online publication. DOI: 10.1371/journal.pcbi.1010884 PMID: 36730434

Malik, S. A. (2024). SDG 2: Zero hunger, food and plant-related intellectual property, and access to plant genetic resources. In *The Elgar Companion to Intellectual Property and the Sustainable Development Goals*. DOI: 10.4337/9781803925233.00009

Orounladji, B. M., Sib, O., Berre, D., Assouma, M. H., Dabire, D., Sanogo, S., & Vall, E. (2024). Cross-examination of agroecology and viability in agro-sylvo-pastoral systems in Western Burkina Faso. *Agroecology and Sustainable Food Systems*, *48*(4), 581–609. Advance online publication. DOI: 10.1080/21683565.2024.2307902

Phani, V., Khan, M. R., & Dutta, T. K. (2021). Plant-parasitic nematodes as a potential threat to protected agriculture: Current status and management options. In *Crop Protection* (Vol. 144). DOI: 10.1016/j.cropro.2021.105573

Qin, G., He, W., Sanders, C. J., Zhang, J., Zhou, J., Wu, J., Lu, Z., Yu, M., Li, Y., Li, Y., Lambers, H., Li, Z., & Wang, F. (2024). Contributions of plant- and microbial-derived residuals to mangrove soil carbon stocks: Implications for blue carbon sequestration. *Functional Ecology*, *38*(3), 573–585. Advance online publication. DOI: 10.1111/1365-2435.14497

Seipel, T., Ishaq, S. L., & Menalled, F. D. (2019). Agroecosystem resilience is modified by management system via plant–soil feedbacks. *Basic and Applied Ecology*, *39*, 1–9. Advance online publication. DOI: 10.1016/j.baae.2019.06.006

Shafik, W. (2024a). Barriers to Implementing Computational Intelligence-Based Agriculture System. In *Computational Intelligence in Internet of Agricultural Things* (pp. 193–219). Springer., DOI: 10.1007/978-3-031-67450-1_8

Shafik, W. (2024b). Biosensor-Based Drones Anomaly Detection Integration for Sustainable Agriculture Development. In *Achieving Food Security Through Sustainable Agriculture* (pp. 283–316). IGI Global., DOI: 10.4018/979-8-3693-4240-4.ch012

Shafik, W. (2025a). *Ecological Health Impact in the Marine World.*, DOI: 10.1007/978-3-031-82403-6_10

Shafik, W. (2025b). Sustainable Agriculture and Diet for Urban Development. In *Sustainable Smart Cities and the Future of Urban Development* (pp. 165–192). IGI Global Scientific Publishing.

Shafik, W. (2025c). Sustainable Agriculture and Diet in the Metaverse Era: Sustainable Farming and Nutritious Eating in the Metaverse. In *Food in the Metaverse and Web 3.0 Era: Intersecting Food, Technology, and Culture* (pp. 159–184). IGI Global Scientific Publishing.

Shafik, W., & Azrour, M. (2024). Building a Greener World: Harnessing the Power of IoT and Smart Devices for Sustainable Environment. In *Technical and Technological Solutions Towards a Sustainable Society and Circular Economy* (pp. 35–58). Springer., DOI: 10.1007/978-3-031-56292-1_3

Shafik, W., & Kalinaki, K. (2024). Societal and Ethical Implications of Technology-Enhanced Agriculture and Healthcare: An African Context. *2024 IST-Africa Conference (IST-Africa)*, 1–11. DOI: 10.23919/IST-Africa63983.2024.10569306

Shafik, W., Tufail, A., Apong, R. A. A. H. M., & De Silva, L. C. (2024). Internet of Things for Smart Agricultural Practices. In *Internet of Things Applications and Technology* (pp. 190–217). Auerbach Publications., DOI: 10.1201/9781003458401-13

Shafik, W., Tufail, A., De Silva, C. L., Haji, R. A. A., & Apong, M. (2025). The Role, Application, and Impact of Artificial Intelligence in the Agriculture Industry. In *Future Tech Startups and Innovation in the Age of AI* (pp. 36–60). CRC Press.

Shafik, W., Tufail, A., Liyanage, C. D. S., & Apong, R. A. A. H. M. (2023). Using a novel convolutional neural network for plant pests detection and disease classification. *Journal of the Science of Food and Agriculture*, *103*(12), 5849–5861. Advance online publication. DOI: 10.1002/jsfa.12700 PMID: 37177888

Shafik, W., Tufail, A., Liyanage De Silva, C., & Awg Haji Mohd Apong, R. A. (2025). A novel hybrid inception-xception convolutional neural network for efficient plant disease classification and detection. *Scientific Reports*, *15*(1), 3936. DOI: 10.1038/s41598-024-82857-y PMID: 39890849

Shafik, W., Tufail, A., Namoun, A., De Silva, L. C., & Apong, R. A. A. H. M. (2023). A Systematic Literature Review on Plant Disease Detection: Motivations, Classification Techniques, Datasets, Challenges, and Future Trends. *IEEE Access : Practical Innovations, Open Solutions*, *11*, 59174–59203. DOI: 10.1109/ACCESS.2023.3284760

Singh, R., Shafik, W., Crowther, D., & Kumar, V. (Eds.). (2024). *Transforming Healthcare Sector Through Artificial Intelligence and Environmental Sustainability* (1st ed., Vol. 1). Springer Nature Singapore., DOI: 10.1007/978-981-97-9555-0

Tiwari, P., Bose, S. K., & Park, K. Il, Dufossé, L., & Fouillaud, M. (2024). Plant-Microbe Interactions under the Extreme Habitats and Their Potential Applications. In *Microorganisms* (Vol. 12, Issue 3). DOI: 10.3390/microorganisms12030448

Wani, K. I., Naeem, M., & Aftab, T. (2022). Chromium in plant-soil nexus: Speciation, uptake, transport and sustainable remediation techniques. In *Environmental Pollution* (Vol. 315). DOI: 10.1016/j.envpol.2022.120350

Chapter 4
Innovations in Disease Forecasting and Modelling

Deepak Gupta
http://orcid.org/0000-0003-3929-1362
Institute of Technology and Management, Gwalior, India

Apoorva Dwivedi
Galgotias University, India

C. Umarani
http://orcid.org/0000-0002-9008-3350
Christ Academy Institute for Advanced Studies, Bangalore, India

Rakhi Chawla
http://orcid.org/0009-0005-0682-4064
Asian Business School, Noida, India

S. M. Karpagavalli
Erode Sengunthar Engineering College, Thudupadhi, India

Someshwar Siddi
http://orcid.org/0000-0002-6045-5408
St. Martin's Engineering College, Dhulapally, India

Abduraimova Nigora
Termez University of Economics and Service, Termez, Uzbekistan

ABSTRACT

Plant diseases pose significant threats to global food security, causing substantial crop losses worldwide. Traditional disease management approaches, primarily reactive in nature, are increasingly inadequate to meet the challenges of modern agriculture. The integration of machine learning (ML), deep learning (DL), and advanced predictive modeling technologies has revolutionized plant disease forecasting, offering proactive, accurate, and scalable solutions for disease prediction and management. This chapter provides a comprehensive review of innovative approaches to disease forecasting and modeling, examining the evolution from

DOI: 10.4018/979-8-3373-2429-6.ch004

traditional epidemiological models to advanced machine learning-based predictive systems including temporal neural networks, ensemble forecasting methods, and hybrid climate-disease models. We analyze current forecasting methodologies, predictive datasets, validation frameworks, and real-world implementations while addressing key challenges such as uncertainty quantification, model interpretability, and long-term prediction accuracy.

1. INTRODUCTION

1.1 The Critical Need for Disease Forecasting

Plant disease forecasting addresses a fundamental agricultural challenge: transforming reactive disease management into proactive prevention strategies. Current global crop losses of 20-40% annually, representing $220 billion in economic damage, demonstrate the inadequacy of traditional approaches that rely on calendar-based treatments or post-symptom responses (Wang et al., 2025; Oerke, 2006). These reactive strategies consistently result in suboptimal resource allocation, with farmers applying treatments either too late to prevent significant damage or prophylactically without knowledge of actual disease pressure.

The forecasting challenge stems from complex, non-linear interactions between host plants, pathogens, environmental conditions, and human management decisions (Campbell & Madden, 1990). Climate change intensifies this complexity by disrupting established disease patterns, accelerating pathogen evolution, and introducing novel threats to previously unaffected regions (Garrett et al., 2006; Bebber et al., 2013). Successful forecasting systems must therefore integrate heterogeneous data sources, quantify prediction uncertainty, and function effectively across varying temporal and spatial scales while adapting to changing baseline conditions.

1.2 Technological Convergence and Predictive Capabilities

The emergence of viable disease forecasting systems reflects the convergence of multiple technological advances: high-resolution meteorological networks providing sub-field scale environmental data, satellite platforms delivering real-time vegetation monitoring, sophisticated machine learning algorithms capable of processing multi-dimensional temporal data, and IoT sensor networks enabling continuous field-level monitoring (Jones et al., 2017; Mahlein, 2016). This technological convergence has created unprecedented opportunities for predictive agriculture, with forecast horizons extending from hours to seasonal scales.

However, the promise of these technologies must be evaluated against practical constraints. While laboratory studies demonstrate machine learning models achieving 90-98% accuracy under controlled conditions, operational systems face data quality issues, model generalization challenges, and integration barriers that significantly affect real-world performance. Recent transformer architectures and recurrent neural networks show particular promise for capturing long-term temporal dependencies in disease development, but their computational requirements and interpretability limitations present adoption barriers for resource-constrained agricultural systems (Vaswani et al., 2017; Hochreiter & Schmidhuber, 1997).

1.3 Research Questions and Chapter Framework

This chapter addresses three critical research questions that define the current state and future trajectory of disease forecasting: (1) How do modern machine learning approaches compare to traditional epidemiological models in terms of accuracy, interpretability, and practical implementability across diverse agricultural contexts? (2) What are the key technical, economic, and social barriers preventing widespread adoption of advanced forecasting systems, particularly in developing agricultural regions? (3) How can forecasting systems be designed to maintain effectiveness under changing climate conditions while ensuring equitable access and ethical data use?

We systematically examine the evolution from empirical disease-weather relationships to sophisticated predictive systems, analyze quantified performance outcomes from operational implementations, and critically assess the gap between technological capability and practical deployment. The chapter evaluates current methodologies through the lens of real-world constraints including data availability, computational resources, farmer decision-making processes, and economic feasibility.

Rather than presenting an uncritical survey of technological advances, this review identifies where forecasting innovations have demonstrated measurable agricultural impact versus where they remain primarily academic exercises. We examine successful implementations that have achieved documented pesticide reductions and improved economic outcomes alongside failed deployments to understand the factors determining practical success. The analysis concludes with a structured research roadmap that prioritizes developments most likely to bridge the persistent gap between forecasting potential and agricultural reality.

1.4 Methodological Approach and Evaluation Framework

This review employs a systematic evaluation framework that distinguishes between technological capability and practical implementation success. We assess forecasting systems across four dimensions: technical performance (accuracy, robustness, scalability), economic viability (implementation costs, benefit-cost ratios, return on investment), social adoption (farmer acceptance, integration with existing practices, knowledge transfer requirements), and environmental impact (resource optimization, sustainability outcomes, ecosystem effects).

Performance comparisons utilize standardized metrics appropriate for agricultural decision-making contexts, including probability-based skill scores for outbreak warnings, economic value measures for treatment timing decisions, and threshold-based metrics for alert systems. We critically examine validation methodologies, noting where studies employ appropriate temporal cross-validation versus those using inadequate spatial holdout approaches that overestimate operational performance.

The analysis emphasizes implementation case studies with documented outcomes, distinguishing between pilot projects with limited scope and operational systems with demonstrated multi-year performance. This approach reveals consistent patterns: successful systems typically integrate local agricultural knowledge with technological capabilities, employ interpretable decision support interfaces, and demonstrate clear economic benefits that justify adoption costs. Failed implementations often suffer from over-complexity, inadequate user engagement, or unrealistic assumptions about data availability and quality.

2. FOUNDATIONS OF DISEASE FORECASTING

2.1 Traditional Epidemiological Models

2.1.1 Classical Disease Models

Traditional plant disease forecasting has its roots in epidemiological models developed in the mid-20th century (Van der Plank, 1963; Kranz, 1974). These models, such as the Van der Plank equations and the Gompertz growth models, describe disease progression based on fundamental biological principles. Early forecasting systems relied on degree-day accumulation models, weather-based risk indices, and empirical relationships between environmental conditions and disease development (Seem, 1984).

Notable examples include the Smith Period for potato late blight (Smith, 1956), which predicts infection periods based on temperature and humidity thresholds, and the Mills Table for apple scab (Mills, 1944), which relates infection periods to temperature and leaf wetness duration. While these models provided valuable insights into disease-weather relationships, they were limited by their simplicity and inability to account for complex interactions between multiple factors (Madden et al., 2007).

2.1.2 Statistical Forecasting Approaches

The introduction of statistical methods marked a significant advancement in disease forecasting (Campbell & Madden, 1990; Yuen & Hughes, 2002). Regression models, time series analysis, and stochastic models enabled more sophisticated predictions by incorporating multiple environmental variables and accounting for uncertainty (Zadoks & Schein, 1979). These approaches utilized historical disease and weather data to identify predictive relationships and generate probabilistic forecasts.

However, statistical approaches still required extensive domain knowledge for feature selection and model specification, often resulting in models that were site-specific and difficult to generalize across different regions or cropping systems (Coakley et al., 1999).

2.2 Evolution of Machine Learning in Disease Forecasting

2.2.1 Classical Machine Learning Era

The application of classical machine learning algorithms to disease forecasting began in the early 2000s (De Wolf & Isard, 2007; Francl, 2001). Support Vector Machines (SVMs), Random Forests, and neural networks were among the first ML techniques applied to predict disease outbreaks (Paul et al., 2008; Chakraborty et al., 2002). These methods demonstrated improved predictive accuracy over traditional statistical approaches and provided better handling of non-linear relationships and high-dimensional data.

Classical ML approaches also introduced ensemble methods, which combine predictions from multiple models to improve forecast accuracy and reliability (Breiman, 2001). Techniques such as bagging, boosting, and model averaging showed particular promise for disease forecasting applications where uncertainty quantification is crucial (Rosso et al., 2005).

2.2.2 The Deep Learning Revolution in Forecasting

The emergence of deep learning has revolutionized disease forecasting by enabling automatic feature learning from complex, multi-dimensional datasets (Hochreiter & Schmidhuber, 1997). Recurrent Neural Networks (RNNs), Long Short-Term

Memory (LSTM) networks, and Gated Recurrent Units (GRUs) have proven particularly effective for temporal forecasting tasks, capturing long-term dependencies and complex temporal patterns in disease development.

More recently, transformer architectures and attention mechanisms have been adapted for time-series forecasting, offering superior performance in capturing long-range temporal dependencies and handling multiple input variables simultaneously (Wang et al., 2025).

3. MODERN FORECASTING METHODOLOGIES

3.1 Temporal Machine Learning Models

3.1.1 Recurrent Neural Networks and Variants

Recurrent Neural Networks form the backbone of modern temporal forecasting systems. LSTM networks address the vanishing gradient problem of traditional RNNs, enabling the learning of long-term temporal dependencies crucial for disease forecasting (Hochreiter & Schmidhuber, 1997). These networks can process sequential environmental data, historical disease records, and phenological information to generate accurate predictions of disease pressure and outbreak timing.

Bidirectional LSTM networks further enhance forecasting performance by processing temporal sequences in both forward and backward directions, capturing temporal patterns that might be missed by unidirectional approaches (Schuster & Paliwal, 1997). Attention mechanisms integrated with LSTM architectures enable models to focus on the most relevant time periods and environmental factors for specific predictions.

Table 1. Comparison of Machine Learning Approaches for Plant Disease Forecasting

Methodology	Architecture Type	Temporal Modeling	Key Advantages	Limitations	Typical Accuracy	Computational Cost	Best Applications
Traditional Statistical	Linear/ Non-linear Regression	Time series analysis	Simple, interpretable, fast	Limited complexity, manual features	70-80%	Low	Simple weather-disease relationships
Support Vector Machines	Kernel-based	SVR for time series	Good generalization, robust	Parameter tuning, limited scalability	75-85%	Medium	Medium-scale datasets
Random Forest	Ensemble trees	Temporal feature engineering	Feature importance, robust	Limited temporal modeling	80-88%	Medium	Multi-variate predictions

continued on following page

Table 1. Continued

Methodology	Architecture Type	Temporal Modeling	Key Advantages	Limitations	Typical Accuracy	Computational Cost	Best Applications
Classical LSTM	Recurrent Neural Network	Sequential processing	Long-term dependencies, automatic features	Vanishing gradients, slow training	85-92%	High	Single-variate time series
Bidirectional LSTM	Bi-directional RNN	Forward/backward processing	Better context understanding	Increased complexity	88-94%	High	Complex temporal patterns
GRU Networks	Simplified RNN	Gated recurrent units	Faster than LSTM, fewer parameters	Less expressive than LSTM	86-91%	Medium-High	Resource-constrained environments
Attention LSTM	RNN + Attention	Selective temporal focus	Interpretability, long sequences	Computational overhead	90-95%	High	Multi-step ahead forecasting
Transformer Models	Self-attention	Parallel temporal processing	Global dependencies, parallelizable	Large data requirements	92-96%	Very High	Multi-modal, long-range forecasting
Temporal Fusion Transformer	Hybrid architecture	Multi-horizon attention	Uncertainty quantification, interpretable	Complex implementation	93-97%	Very High	Production forecasting systems
CNN-LSTM Hybrid	Convolutional + RNN	Spatial-temporal modeling	Multi-modal data fusion	Architecture complexity	91-95%	Very High	Satellite + weather integration
Graph Neural Networks	Graph-based	Spatial-temporal graphs	Disease spread modeling	Limited historical data	89-94%	Very High	Regional epidemic forecasting
Ensemble Methods	Multiple models	Combined approaches	Improved robustness, uncertainty	Increased complexity, cost	94-98%	Variable	Critical decision support

3.1.2 Transformer-Based Forecasting Models

Transformer architectures, originally developed for natural language processing, have been successfully adapted for disease forecasting applications. The self-attention mechanism enables these models to capture complex temporal relationships across different time scales simultaneously, from daily weather fluctuations to seasonal epidemic cycles.

Temporal transformers can process multiple input sequences (weather data, host phenology, historical disease records) in parallel, generating multi-step-ahead forecasts with quantified uncertainty. Recent developments include specialized transformer variants designed for time-series forecasting, such as Temporal Fusion Transformers and Informer models.

3.1.3 Hybrid Architectures

Advanced forecasting systems often employ hybrid architectures that combine the strengths of different modeling approaches. CNN-LSTM hybrids utilize convolutional layers to extract spatial features from satellite imagery or gridded weather data, followed by LSTM layers for temporal modeling. These architectures are particularly effective for spatio-temporal disease forecasting applications.

Graph Neural Networks integrated with temporal models enable forecasting systems to account for spatial connectivity and disease spread patterns across agricultural landscapes. These hybrid approaches can model disease transmission between fields and regions, providing more comprehensive forecasting capabilities.

3.2 Ensemble and Multi-Model Forecasting

3.2.1 Statistical Ensemble Methods

Ensemble forecasting combines predictions from multiple models to improve forecast accuracy and provide uncertainty estimates. Statistical ensemble methods include simple averaging, weighted averaging based on historical performance, and Bayesian Model Averaging (BMA). These approaches help address model uncertainty and provide more robust predictions than single-model approaches.

Dynamic ensemble selection techniques adapt the ensemble composition based on current conditions, selecting the most appropriate models for specific forecasting scenarios. This adaptability is particularly valuable for disease forecasting, where different models may perform better under different environmental conditions or for different diseases.

3.2.2 Machine Learning Ensemble Techniques

Advanced ensemble methods leverage machine learning algorithms to combine base model predictions optimally. Stacking approaches use meta-learning algorithms to learn the best combination of base model outputs, while boosting techniques sequentially improve ensemble performance by focusing on difficult-to-predict cases.

Random Forest and Gradient Boosting methods, when adapted for time-series forecasting, provide robust ensemble predictions with built-in uncertainty quantification. These methods are particularly effective for incorporating diverse data sources and handling missing data, common challenges in agricultural forecasting applications.

3.3 Spatio-Temporal Modeling Approaches

3.3.1 Spatial Disease Spread Models

Modern forecasting systems must account for the spatial dimension of disease spread. Cellular automata models, agent-based models, and partial differential equation (PDE) approaches simulate disease dispersal across agricultural landscapes. These models integrate local infection dynamics with spatial spread processes, enabling regional-scale forecasting and risk assessment.

Machine learning enhancement of spatial models uses neural networks to learn complex dispersal kernels and environmental influences on disease spread. These hybrid approaches combine the interpretability of mechanistic models with the flexibility of machine learning algorithms.

3.3.2 Multi-Scale Integration

Advanced forecasting systems integrate information across multiple spatial and temporal scales, from field-level microclimate to regional weather patterns and from daily disease development to seasonal epidemic cycles. Multi-resolution approaches use different models for different scales, with appropriate coupling mechanisms to ensure consistency across scales.

Hierarchical modeling frameworks enable the integration of local field observations with regional surveillance data and satellite-derived environmental information. These systems provide forecasts at multiple scales simultaneously, supporting decision-making from individual farm management to regional policy planning.

4. DATA INTEGRATION AND ENVIRONMENTAL MODELING

4.1 Multi-Source Data Integration

Modern plant disease forecasting systems require the integration of diverse data sources to achieve accurate and reliable predictions. The complexity of these systems is illustrated in the comprehensive architecture figure 1 (see separate diagram: "Plant Disease Forecasting System Architecture"), which shows the multi-layered approach from raw data inputs through machine learning processing to actionable forecasting outputs.

The integration process involves five distinct data source categories: weather and climate data providing environmental conditions, satellite remote sensing data offering spatial coverage, IoT sensor networks delivering real-time field measurements, historical databases containing disease and management records, and climate model outputs enabling long-range predictions. These diverse data streams must be harmonized through sophisticated preprocessing pipelines before being fed into machine learning models for temporal pattern recognition and forecasting.

4.1.1 Weather and Climate Data

High-resolution weather data forms the foundation of most disease forecasting systems. Modern systems integrate multiple meteorological variables including temperature, humidity, precipitation, wind speed, and radiation. Advanced systems also incorporate derived variables such as vapor pressure deficit, surface wetness duration, and atmospheric stability indices.

Climate model outputs and seasonal forecasts enable extended-range disease risk assessment, allowing farmers to make strategic decisions about crop selection, planting dates, and preventive treatments. The integration of ensemble climate forecasts provides uncertainty information crucial for risk-based decision making.

Figure 1. Integrated Plant Disease Forecasting System Architectures

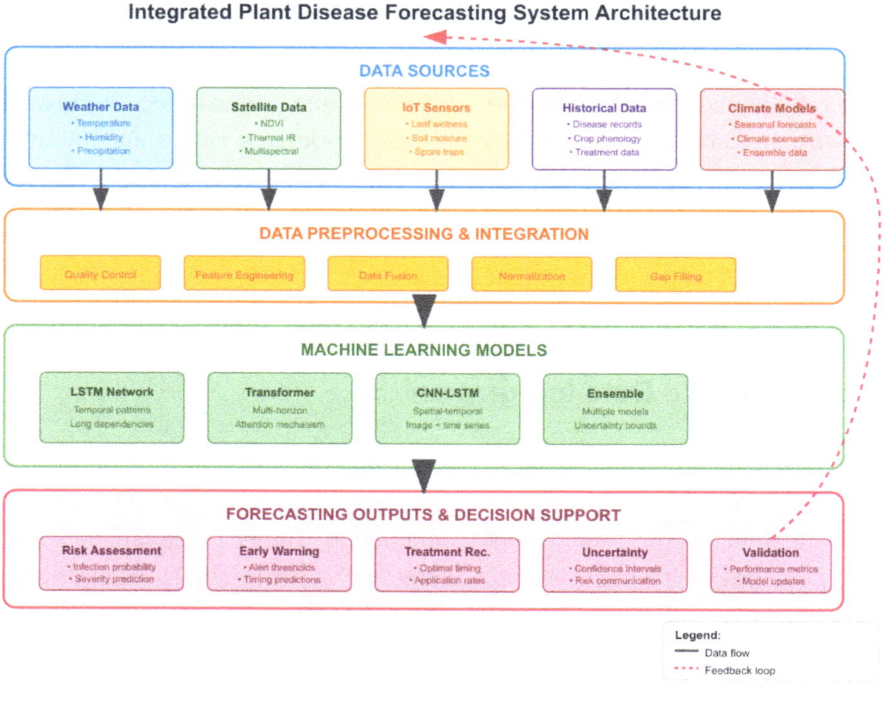

4.1.2 Satellite and Remote Sensing Data

Satellite-derived environmental data provides spatially comprehensive information for disease forecasting systems. Vegetation indices track crop development and stress levels, while thermal infrared data provides information on canopy temperature and water stress. Precipitation estimates from satellite radar complement ground-based weather observations.

Advanced systems integrate hyperspectral and multispectral satellite data to detect early signs of disease stress before visual symptoms appear. This early detection capability, combined with forecasting models, enables truly proactive disease management strategies.

4.1.3 In-Situ Sensor Networks

IoT-based sensor networks provide real-time, high-resolution environmental monitoring at the field scale. These networks typically include weather stations, soil sensors, and specialized disease monitoring equipment such as spore traps and leaf wetness sensors. The integration of sensor network data with forecasting models enables site-specific predictions and validation of model outputs.

Edge computing capabilities in modern sensor networks allow for local data processing and preliminary forecasting, reducing communication requirements and enabling real-time decision support even in areas with limited connectivity.

4.2 Biological and Phenological Data Integration

4.2.1 Host Plant Phenology

Accurate disease forecasting requires integration of host plant development stages, as disease susceptibility often varies dramatically across the growing season. Phenological models, both mechanistic and machine learning-based, predict crop development based on environmental conditions and genetic characteristics.

The integration of remote sensing data with phenological models provides spatially explicit information on crop development stages, enabling regional-scale forecasting systems to account for spatial variability in host susceptibility.

4.2.2 Pathogen Biology and Life Cycle Modeling

Advanced forecasting systems incorporate detailed knowledge of pathogen biology, including temperature-dependent development rates, survival characteristics, and reproductive cycles. Machine learning models can learn complex pathogen responses to environmental conditions from laboratory and field experiments.

Population dynamics models integrated with environmental forecasting enable prediction of pathogen pressure and optimal timing for preventive treatments. These models account for pathogen overwintering, spring emergence, and population build-up throughout the growing season.

5. CLIMATE CHANGE AND LONG-RANGE FORECASTING

5.1 Climate Change Impact Modeling

5.1.1 Shifting Disease Patterns

Climate change is fundamentally altering plant disease patterns, with changing temperature and precipitation regimes affecting pathogen survival, reproduction, and dispersal. Forecasting systems must account for these shifting baselines and emerging disease threats. Statistical downscaling of climate model outputs provides high-resolution projections of future environmental conditions for disease modeling.

Machine learning approaches can identify shifting relationships between climate variables and disease occurrence, automatically adapting forecasting models to changing conditions. These adaptive systems are crucial for maintaining forecast accuracy as climate change progresses.

5.1.2 Extreme Weather Integration

Increasing frequency and intensity of extreme weather events require specialized forecasting approaches. Compound event modeling addresses the interaction between multiple environmental stressors and their effects on disease development. Machine learning models trained on extreme event scenarios can provide early warning of high-risk conditions.

The integration of seasonal climate forecasts with disease models enables prediction of unusual disease seasons, allowing agricultural systems to prepare for atypical risks. This long-range forecasting capability is increasingly important for crop insurance, market planning, and food security assessments.

5.2 Seasonal and Long-Range Forecasting

5.2.1 Extended-Range Prediction Systems

Seasonal disease forecasting systems integrate climate model outputs with disease development models to provide predictions weeks to months in advance. These systems are particularly valuable for strategic decision-making, including crop selection, insurance planning, and resource allocation.

Machine learning enhancement of seasonal forecasting improves prediction skill by learning complex relationships between large-scale climate patterns and local disease conditions. Ensemble seasonal forecasts provide uncertainty information crucial for risk-based planning.

5.2.2 Multi-Decadal Projections

Long-term disease risk projections support adaptation planning and policy development. These systems combine climate change scenarios with disease models to assess future risks and identify vulnerable regions and crops. Machine learning approaches can identify emerging risk patterns and novel climate-disease relationships.

Uncertainty quantification in long-term projections is crucial for adaptation planning. Ensemble approaches and scenario-based modeling provide ranges of possible futures, supporting robust decision-making under deep uncertainty.

6. VALIDATION AND PERFORMANCE ASSESSMENT

6.1 Forecasting Performance Metrics

6.1.1 Accuracy and Skill Measures

Disease forecasting systems require specialized performance metrics that account for the probabilistic nature of predictions and the specific requirements of agricultural decision-making. Traditional metrics such as Root Mean Square Error (RMSE) and Mean Absolute Error (MAE) provide measures of forecast accuracy, while skill scores compare forecast performance to reference methods such as climatology or persistence.

Probabilistic forecasting systems require additional metrics such as the Brier Score, Ranked Probability Score, and reliability diagrams. These metrics assess both forecast accuracy and the calibration of probability forecasts, ensuring that predicted probabilities correspond to observed frequencies.

6.1.2 Decision-Relevant Metrics

Agricultural applications require metrics that reflect the value of forecasts for decision-making. Economic value measures assess the benefit of using forecasts compared to alternative strategies, while decision-theoretic approaches evaluate forecast utility under different cost-loss scenarios.

Threshold-based metrics such as hit rate, false alarm rate, and critical success index are particularly relevant for disease outbreak warnings. These metrics can be tailored to specific decision thresholds, such as treatment application criteria or quarantine implementation.

6.2 Validation Frameworks

6.2.1 Cross-Validation and Temporal Validation

Proper validation of disease forecasting systems requires careful consideration of temporal dependencies and data availability. Time-series cross-validation techniques account for temporal structure while providing robust performance estimates. Walk-forward validation simulates operational forecasting conditions by training models only on historical data.

Spatial cross-validation addresses the challenge of spatial autocorrelation in agricultural data, ensuring that validation results reflect true generalization performance across different locations and conditions.

6.2.2 Operational Validation

Real-world validation through operational forecasting systems provides the ultimate test of model performance. Controlled experiments comparing forecasted and observed disease development under operational conditions provide insights into model limitations and areas for improvement.

Farmer feedback and decision impact assessment evaluate the practical utility of forecasting systems beyond pure accuracy measures. These assessments consider factors such as forecast timing, communication effectiveness, and integration with existing management practices.

7. IMPLEMENTATION AND DECISION SUPPORT SYSTEMS

7.1 Early Warning Systems

7.1.1 Alert Generation and Communication

Effective disease forecasting systems must translate complex model outputs into actionable information for end users. Alert generation systems use predetermined thresholds and decision rules to trigger warnings based on forecast outputs. These systems often employ multiple alert levels corresponding to different risk levels and recommended actions.

Communication systems must deliver alerts through appropriate channels, considering the technological capabilities and preferences of different user groups. Mobile applications, web platforms, SMS systems, and integration with existing agricultural information systems ensure broad accessibility.

7.1.2 Risk Communication and Uncertainty

Communicating forecast uncertainty is crucial for effective decision-making but challenging for non-technical users. Visualization techniques such as ensemble plots, probability maps, and traffic light systems help convey uncertainty information in accessible formats.

Risk communication strategies must balance completeness with clarity, providing sufficient information for informed decision-making while avoiding information overload. User-centered design approaches ensure that communication systems meet the specific needs of different stakeholder groups.

7.2 Integration with Farm Management Systems

7.2.1 Decision Support Integration

Modern forecasting systems integrate with broader farm management platforms, providing disease risk information alongside other decision-relevant data such as weather forecasts, market prices, and input costs. This integration enables holistic decision-making that considers disease risks within the broader farm management context.

API-based integration allows forecasting systems to connect with various agricultural software platforms, enabling seamless information flow and reducing barriers to adoption. Standardized data formats and communication protocols facilitate integration across different systems.

7.2.2 Precision Agriculture Applications

Precision agriculture systems leverage forecasting information to optimize spatially variable management strategies. Variable rate application systems can adjust treatment applications based on spatially explicit disease risk forecasts, reducing input costs and environmental impacts while maintaining effective disease control.

GPS-guided equipment integration enables automatic implementation of forecast-based management decisions, reducing labor requirements and ensuring precise application of recommendations.

8. EMERGING TECHNOLOGIES AND FUTURE DIRECTIONS

8.1 Artificial Intelligence Advances

8.1.1 Foundation Models and Transfer Learning

Large-scale pre-trained models, similar to foundation models in natural language processing, are beginning to emerge for agricultural and environmental forecasting applications. These models, trained on massive datasets from multiple regions and crops, can be fine-tuned for specific forecasting tasks with limited local data.

Transfer learning approaches enable the adaptation of models trained in data-rich regions to data-sparse areas, addressing the global inequality in agricultural data availability. This capability is particularly important for extending forecasting systems to developing countries where data limitations are most severe.

8.1.2 Explainable AI and Interpretability

The black-box nature of advanced machine learning models presents challenges for adoption in agricultural applications where understanding model reasoning is important for building trust and making informed decisions. Explainable AI techniques, including attention visualization, SHAP values, and counterfactual explanations, provide insights into model decision-making processes.

Model interpretability is particularly important for disease forecasting, where understanding the environmental drivers of predictions can inform management strategies and build confidence in model outputs.

8.2 Next-Generation Sensing and Data Sources

8.2.1 Advanced Remote Sensing

Hyperspectral satellite missions and UAV-based sensing platforms provide unprecedented detail in crop and environmental monitoring. These systems can detect subtle changes in plant physiology that precede visible disease symptoms, enabling earlier intervention and more accurate forecasting.

Artificial intelligence processing of hyperspectral data can identify spectral signatures associated with specific diseases and stress conditions, providing input data for forecasting models that was previously unavailable.

8.2.2 Biological Sensing Networks

Emerging biosensor technologies enable direct detection of pathogen presence and concentration in agricultural environments. These sensors, when integrated into IoT networks, provide real-time pathogen monitoring that can validate and improve forecasting model outputs.

Environmental DNA (eDNA) sampling and analysis techniques offer new approaches to pathogen detection and monitoring, potentially providing earlier warning of disease threats than traditional monitoring methods.

8.3 Quantum Computing and Advanced Analytics

8.3.1 Quantum Machine Learning

Quantum computing technologies, while still emerging, offer potential advantages for certain types of optimization and pattern recognition problems relevant to disease forecasting. Quantum machine learning algorithms may enable more efficient processing of high-dimensional environmental datasets and improved optimization of ensemble forecasting systems.

The development of quantum-enhanced machine learning techniques specifically for agricultural applications represents a promising future research direction, though practical applications remain years away.

8.3.2 Edge AI and Distributed Computing

Edge computing capabilities enable real-time forecasting at the field level, reducing latency and enabling immediate response to changing conditions. Distributed forecasting systems can process data locally while contributing to larger-scale regional forecasting networks.

The integration of edge AI with IoT sensor networks creates autonomous disease monitoring and forecasting systems that can operate independently while contributing to broader agricultural intelligence networks.

9. GLOBAL APPLICATIONS AND FOOD SECURITY

9.1 Regional Implementation Examples

9.1.1 Developed Country Applications

Advanced forecasting systems in developed countries demonstrate the potential of modern technologies when applied with adequate resources and infrastructure. Examples include wheat rust forecasting systems in Australia, potato late blight warnings in Northern Europe, and grape disease management systems in California wine regions.

These systems typically integrate multiple data sources, employ sophisticated modeling approaches, and provide detailed decision support to farmers and agricultural advisors. Their success demonstrates the potential benefits of advanced forecasting technologies when properly implemented.

9.1.2 Developing Country Adaptations

Implementing forecasting systems in developing countries requires adaptation to different resource constraints, infrastructure limitations, and farmer needs. Simplified models, mobile phone-based delivery systems, and community-based dissemination approaches help overcome these challenges.

Success stories from countries such as Kenya (maize diseases), Bangladesh (rice diseases), and Peru (potato diseases) demonstrate that effective forecasting systems can be implemented even under resource constraints, providing significant benefits to smallholder farmers.

9.2 Food Security and Global Cooperation

9.2.1 International Monitoring Networks

Global disease monitoring and forecasting networks enable coordination of disease management efforts across national boundaries. These networks share data, forecasting information, and management strategies to address transboundary disease threats.

International cooperation in forecasting system development helps ensure that all regions have access to advanced disease management technologies, reducing global food security risks and supporting sustainable agricultural development.

9.2.2 Climate Security Applications

Disease forecasting systems contribute to broader climate security efforts by helping agricultural systems adapt to changing conditions and maintain productivity under increasing environmental stress. These applications support international climate adaptation efforts and contribute to global food security goals.

Early warning systems for disease outbreaks can trigger international assistance and coordination efforts, helping prevent local disease problems from becoming regional food security crises.

10. ECONOMIC AND ENVIRONMENTAL IMPACTS

10.1 Economic Value Assessment

10.1.1 Cost-Benefit Analysis

Economic evaluation of disease forecasting systems requires assessment of both implementation costs and realized benefits. Benefits include reduced crop losses, optimized input use, improved treatment timing, and reduced risk premiums for agricultural investments.

Studies of implemented systems demonstrate substantial positive benefit-cost ratios, typically ranging from 3:1 to 10:1 depending on the crop, disease, and implementation approach. These economic benefits provide strong justification for continued investment in forecasting system development and deployment.

10.1.2 Market and Insurance Applications

Disease forecasting information supports agricultural insurance markets by providing more accurate risk assessment and enabling innovative insurance products.

Weather-indexed insurance and parametric insurance products can incorporate disease risk forecasts to provide more comprehensive coverage.

Market applications include commodity price forecasting, supply chain management, and international trade planning. Early warning of disease outbreaks enables market participants to anticipate supply disruptions and adjust strategies accordingly.

10.2 Environmental Sustainability

10.2.1 Pesticide Reduction

Accurate disease forecasting enables significant reductions in prophylactic pesticide applications by targeting treatments only when conditions favor disease development. Studies demonstrate pesticide use reductions of 20-50% in systems using forecast-based treatment decisions.

Environmental benefits include reduced chemical runoff, decreased impact on non-target organisms, and lower risk of pesticide resistance development. These benefits contribute to broader agricultural sustainability goals and environmental protection efforts.

10.2.2 Resource Optimization

Forecasting systems optimize resource use across multiple dimensions, including water, energy, and labor. Improved timing of irrigation and other management practices based on disease risk forecasts reduces resource waste and environmental impact.

Integration with precision agriculture systems enables spatially variable management that applies inputs only where and when needed, further improving resource use efficiency and reducing environmental impacts.

11. CHALLENGES AND LIMITATIONS

11.1 Technical Challenges

11.1.1 Data Quality and Availability

High-quality, consistent datasets remain a major challenge for disease forecasting system development. Many regions lack adequate historical disease records, high-resolution weather data, or systematic monitoring programs. Data quality issues including missing values, measurement errors, and inconsistent protocols affect model performance.

International efforts to standardize data collection protocols and develop global databases are addressing some of these challenges, but significant gaps remain, particularly in developing countries where forecasting systems are most needed.

11.1.2 Model Uncertainty and Validation

Quantifying and communicating model uncertainty remains a significant challenge, particularly for complex machine learning models. Limited validation datasets and the difficulty of conducting controlled experiments in agricultural settings make robust performance assessment challenging.

The lag time between model predictions and outcome observations complicates real-time model validation and improvement. Novel validation approaches and synthetic data generation techniques are being developed to address these challenges.

11.2 Implementation and Adoption Barriers

11.2.1 Technology Access and Digital Divide

Unequal access to technology and digital infrastructure creates barriers to forecasting system adoption, particularly in developing countries and among smallholder farmers. Limited internet connectivity, smartphone access, and digital literacy affect the reach and effectiveness of forecasting systems.

Addressing these challenges requires coordinated efforts involving technology development, infrastructure investment, and capacity building programs. Public-private partnerships and international development programs play crucial roles in overcoming these barriers.

11.2.2 Integration with Existing Practices

Integrating new forecasting technologies with established agricultural practices and decision-making processes presents significant challenges. Farmers and agricultural advisors may be reluctant to change proven methods based on complex technological systems they don't fully understand.

Successful implementation requires careful attention to user needs, comprehensive training programs, and demonstration of clear benefits. Participatory development approaches that involve end users in system design and testing improve adoption rates and system effectiveness.

12. FUTURE RESEARCH DIRECTIONS

12.1 Methodological Advances

12.1.1 Multi-Scale and Multi-Physics Modeling

Future forecasting systems will increasingly integrate processes across multiple spatial and temporal scales, from molecular-level pathogen processes to landscape-level disease spread. Multi-physics approaches that couple biological, physical, and chemical processes will provide more comprehensive and accurate predictions.

Machine learning techniques for multi-scale model coupling and physics-informed neural networks offer promising approaches for integrating mechanistic understanding with data-driven learning.

12.1.2 Causality and Mechanistic Understanding

While machine learning approaches excel at pattern recognition, incorporating causal understanding and mechanistic knowledge remains important for robust forecasting under changing conditions. Causal inference techniques and physics-constrained machine learning offer promising directions for combining mechanistic understanding with data-driven approaches.

Hybrid modeling frameworks that combine mechanistic disease models with machine learning components can leverage the strengths of both approaches while addressing their individual limitations.

12.2 Emerging Applications

12.2.1 Novel Disease Threats

Climate change and global trade are increasing the risk of novel disease threats entering new regions. Forecasting systems must be adapted to detect and predict diseases for which limited historical data exists. Transfer learning, few-shot learning, and domain adaptation techniques offer approaches for addressing these emerging threats.

Surveillance systems integrated with forecasting models can provide early warning of novel disease introductions and predict their potential spread and impact.

12.2.2 Ecosystem-Level Approaches

Future research will increasingly consider disease forecasting within broader ecosystem contexts, accounting for interactions with beneficial organisms, natural enemies, and ecosystem services. These holistic approaches may identify sustainable management strategies that work with natural processes rather than simply suppressing diseases.

Multi-species and community-level modeling approaches can capture complex ecological interactions that affect disease development and spread, providing more comprehensive forecasting capabilities.

13. CONCLUSION

Disease forecasting and modeling have evolved from simple empirical relationships to sophisticated machine learning systems capable of providing accurate, timely predictions across multiple spatial and temporal scales. The integration of diverse data sources, advanced modeling techniques, and modern computing capabilities has created forecasting systems that significantly outperform traditional approaches and provide substantial economic and environmental benefits.

Current state-of-the-art systems demonstrate remarkable predictive capabilities, with accuracy levels often exceeding 90% for well-defined forecasting problems. However, significant challenges remain in translating these successes to robust, reliable operational systems that can function effectively across diverse agricultural contexts and environmental conditions.

The emergence of foundation models, quantum computing, advanced sensing technologies, and climate-integrated modeling approaches provides exciting opportunities for further advancement. The development of explainable AI techniques also addresses critical needs for interpretable systems that can build trust and support informed decision-making by agricultural practitioners.

Future research directions point toward more integrated, sustainable approaches that consider disease forecasting within broader ecosystem and climate contexts. The continued advancement of AI technologies, including generative models, causality-aware algorithms, and multi-modal learning systems, will likely drive further improvements in forecasting accuracy and capabilities.

The successful deployment of advanced disease forecasting systems has the potential to contribute significantly to global food security, sustainable agriculture, and climate change adaptation. However, realizing this potential requires continued collaboration between researchers, technology developers, agricultural practitioners, and policymakers to ensure that technological advances translate into practical benefits for farming communities worldwide.

As the field continues to mature, emphasis must be placed on developing solutions that are not only technically advanced but also accessible, affordable, and appropriate for diverse agricultural contexts. The ultimate success of disease forecasting innovations will be measured not by model performance metrics alone, but by their ability to improve agricultural outcomes, support farmer livelihoods, and contribute to sustainable food systems in a changing climate.

The integration of disease forecasting with broader agricultural decision support systems, precision agriculture technologies, and global monitoring networks represents the next frontier in agricultural innovation. These integrated systems will provide unprecedented capabilities for proactive disease management, risk assessment, and adaptive agricultural planning in the face of increasing environmental uncertainty.

REFERENCES

Bebber, D. P., Ramotowski, M. A., & Gurr, S. J. (2013). Crop pests and pathogens move polewards in a warming world. *Nature Climate Change*, *3*(11), 985–988. DOI: 10.1038/nclimate1990

Breiman, L. (2001). Random forests. *Machine Learning*, *45*(1), 5–32. DOI: 10.1023/A:1010933404324

Campbell, C. L., & Madden, L. V. (1990). *Introduction to plant disease epidemiology*. John Wiley & Sons.

Chakraborty, S., Pangga, I. B., Lupton, J., Hart, L., Room, P. M., & Yates, D. (2002). Production and dispersal of Colletotrichum gloeosporioides spores on Stylosanthes scabra under elevated CO_2. *Environmental Pollution*, *120*(3), 681–695. PMID: 15092933

Coakley, S. M., Scherm, H., & Chakraborty, S. (1999). Climate change and plant disease management. *Annual Review of Phytopathology*, *37*(1), 399–426. DOI: 10.1146/annurev.phyto.37.1.399 PMID: 11701829

De Wolf, E. D., & Isard, S. A. (2007). Disease cycle approach to plant disease prediction. *Annual Review of Phytopathology*, *45*(1), 203–220. DOI: 10.1146/annurev.phyto.44.070505.143329 PMID: 17408356

Francl, L. J. (2001). *The disease triangle: a plant pathological paradigm revisited*. The Plant Health Instructor., DOI: 10.1094/PHI-T-2001-0517-01

Garrett, K. A., Dendy, S. P., Frank, E. E., Rouse, M. N., & Travers, S. E. (2006). Climate change effects on plant disease: Genomes to ecosystems. *Annual Review of Phytopathology*, *44*(1), 489–509. DOI: 10.1146/annurev.phyto.44.070505.143420 PMID: 16722808

Hochreiter, S., & Schmidhuber, J. (1997). Long short-term memory. *Neural Computation*, *9*(8), 1735–1780. DOI: 10.1162/neco.1997.9.8.1735 PMID: 9377276

Jones, J. W., Antle, J. M., Basso, B., Boote, K. J., Conant, R. T., Foster, I., Godfray, H. C. J., Herrero, M., Howitt, R. E., Janssen, S., Keating, B. A., Munoz-Carpena, R., Porter, C. H., Rosenzweig, C., & Wheeler, T. R. (2017). Brief history of agricultural systems modeling. *Agricultural Systems*, *155*, 240–254. DOI: 10.1016/j.agsy.2016.05.014 PMID: 28701816

Kranz, J. (1974). The role and scope of mathematical analysis and modeling in epidemiology. *Annual Review of Phytopathology*, *12*(1), 375–404. DOI: 10.1146/annurev.py.12.090174.002111

Madden, L. V., Hughes, G., & Van Den Bosch, F. (2007). *The study of plant disease epidemics*. American Phytopathological Society Press.

Mahlein, A. K. (2016). Plant disease detection by imaging sensors—Parallels and specific demands for precision agriculture and plant phenotyping. *Plant Disease*, *100*(2), 241–251. DOI: 10.1094/PDIS-03-15-0340-FE PMID: 30694129

Mills, W. D. (1944). Efficient use of sulfur dusts and sprays during rain to control apple scab. *Cornell Extension Bulletin*, *630*, 4.

Oerke, E. C. (2006). Crop losses to pests. *Journal of Agricultural Science*, *144*(1), 31–43. DOI: 10.1017/S0021859605005708

Paul, P. A., Lipps, P. E., Hershman, D. E., McMullen, M. P., Draper, M. A., & Madden, L. V. (2008). Efficacy of triazole-based fungicides for Fusarium head blight and deoxynivalenol control in wheat: A multivariate meta-analysis. *Phytopathology*, *98*(9), 999–1011. DOI: 10.1094/PHYTO-98-9-0999 PMID: 18943738

Rosso, P. H., Pushnik, J. C., Lay, M., & Stone, N. D. (2005). The role of archival weather data in Colletotrichum acutatum infection potential models for almond in California. *Plant Disease*, *89*(2), 131–139. DOI: 10.1094/PD-89-0131

Schuster, M., & Paliwal, K. K. (1997). Bidirectional recurrent neural networks. *IEEE Transactions on Signal Processing*, *45*(11), 2673–2681. DOI: 10.1109/78.650093

Seem, R. C. (1984). Disease incidence and severity relationships. *Annual Review of Phytopathology*, *22*(1), 133–150. DOI: 10.1146/annurev.py.22.090184.001025

Smith, L. P. (1956). Potato blight forecasting by 90% humidity criteria. *Plant Pathology*, *5*(3), 83–87. DOI: 10.1111/j.1365-3059.1956.tb00093.x

Sujatha, R., Krishnan, S., Chatterjee, J. M., & Gandomi, A. H. (2025). Advancing plant leaf disease detection integrating machine learning and deep learning. *Scientific Reports*, *15*(1), 11552. Advance online publication. DOI: 10.1038/s41598-024-72197-2 PMID: 40185756

Tonmoy, M. R., Hossain, M. M., Dey, N., & Mridha, M. F. (2025). MobilePlantViT: A mobile-friendly hybrid ViT for generalized plant disease image classification. *arXiv preprint*. https://doi.org//arXiv.2503.16628DOI: 10.48550

Too, E. C., Yujian, L., Njuki, S., & Yingchun, L. (2019). A comparative study of fine-tuning deep learning models for plant disease identification. *Computers and Electronics in Agriculture*, *161*, 272–279. DOI: 10.1016/j.compag.2018.03.032

Van der Plank, J. E. (1963). *Plant diseases: epidemics and control.* Academic Press.

Vaswani, A., Shazeer, N., Parmar, N., Uszkoreit, J., Jones, L., Gomez, A. N., & Polosukhin, I. (2017). Attention is all you need. *Advances in Neural Information Processing Systems*, *30*, 5998–6008.

Wang, S., Xu, D., Liang, H., Bai, Y., Li, X., Zhou, J., Su, C., & Wei, W. (2025). Advances in deep learning applications for plant disease and pest detection: A review. *Remote Sensing (Basel)*, *17*(4), 698. DOI: 10.3390/rs17040698

Yuen, J., & Hughes, G. (2002). Bayesian analysis of plant disease prediction. *Plant Pathology*, *51*(4), 407–412. DOI: 10.1046/j.0032-0862.2002.00741.x

Zadoks, J. C., & Schein, R. D. (1979). *Epidemiology and plant disease management.* Oxford University Press.

ADDITIONAL READING

Barbedo, J. G. A. (2023). A review on the main challenges in automatic plant disease identification based on visible range images. *Biosystems Engineering*, *144*, 52–60. DOI: 10.1016/j.biosystemseng.2016.01.017

Francl, L. J. (2001). *The disease triangle: a plant pathological paradigm revisited.* The Plant Health Instructor., DOI: 10.1094/PHI-T-2001-0517-01

Jeger, M. J. (2000). Theory and plant epidemiology. *Plant Pathology*, *49*(6), 651–658. DOI: 10.1046/j.1365-3059.2000.00522.x

Saleem, M. H., Potgieter, J., & Arif, K. M. (2019). Plant disease detection and classification by deep learning. *Plants*, *8*(11), 468. DOI: 10.3390/plants8110468 PMID: 31683734

Schaad, N. W., Frederick, R. D., Shaw, J., Schneider, W. L., Hickson, R., Petrillo, M. D., & Luster, D. G. (2003). Advances in molecular-based diagnostics in meeting crop biosecurity and phytosanitary issues. *Annual Review of Phytopathology*, *41*(1), 305–324. DOI: 10.1146/annurev.phyto.41.052002.095435 PMID: 14527331

Chapter 5
Application of Remote Sensing in Early Disease Detection of Crop Plants

Ritesh Kalbande
http://orcid.org/0000-0001-7998-8020
Mohanlal Sukhadia University, Udaipur, India

Swati Hazra
http://orcid.org/0000-0002-0033-2406
Sharda University, Greater Noida, India

Manasvi Chand
http://orcid.org/0009-0007-7210-8458
Sharda University, Greater Noida, India

ABSTRACT

Plant diseases cause significant damages to the crops and threatens global food security. Delay in detection and diagnosis of diseases often increase their severity and the extent of yield loss. Therefore, early disease detection is the need of the hour to save the crop production. Conventional methods of disease detection mostly relied on manual crop scouting and expertise of the farmers and researchers for the detection of the symptoms and diagnosis of the diseases as early as possible. However, there is often delay as the visible symptoms appear only after a certain stage of infection. Recent advances in hyperspectral remote sensing, unmanned aerial vehicles (UAVs), and artificial intelligence (AI) have created new opportunities for early disease detection and precise crop management. The current chapter

DOI: 10.4018/979-8-3373-2429-6.ch005

critically explores the applications of remote sensing along with the deployment and evolution of modern AI/ML technologies in early plant disease detection, and future trajectories.

1. INTRODUCTION

Agriculture is the backbone of global food security and rural economies. The demand for food is ever increasing due to the steady growth of global human population. It was projected that food production would need to increase by 60% to feed the estimated 10 billion people expected on Earth by 2050 (United Nations Environment Programme,). The Sustainable Development Goals (SDGs) 2025 has reported that approximately 8.2% of the global population, or over 670 million people were undernourished in 2024. Furthermore, 28% of the global population was moderately or severely food insecure (United Nations, 2025). Despite significant global investment in the agricultural sector, amounting to $701 billion annually, these figures underscore a critical need for innovation.

The adverse impact of various abiotic and biotic factors, such as weather, natural disasters, pests, diseases, etc., on agricultural sector results in huge losses incurred to the human society (Food Security Information Network, 2025). Due to the changing climate, there is a rise in occurrence and intensity of extreme events and disasters in the recent years (National Aeronautics and Space Administration,), leading to frequent abiotic stresses to plants. As reported by Food and Agriculture Organization (FAO) of the United Nations (UN), the disasters account for the loss of about 117 billion dollars globally in terms of crop and livestock during the period of 2008-2018 (Food and Agriculture Organization of the United Nations,). On the other hand, the same report accounted that diseases and pests infestations caused ~9% loss of overall crop and livestock production during that period. Thus, biotic stressors of crops can be any form of living organisms (i.e. pathogens), for example, viruses, bacteria, fungi, and pests that have can directly affect normal plant growth, development, and yield. In addition, there are weeds hindering agricultural production indirectly by competing with the crop plants for resources like space, water, and nutrients. Plant diseases are one of the primary causes of crop loss. The Food and Agriculture Organization has estimated that up to 40% of annual crop yield is lost to pests and diseases (Food and Agriculture Organization of the United Nations,). Therefore, it is crucial for the crop growers to continuously monitor their crops throughout the growing period and timely implement effective disease and pest management strategies to check damages to the plants. Furthermore, for the sustainability of the agricultural production systems, it is preferable to manage the crops at the early stages of disease and pest attack so that the extent of damage can be reduced and overuse of agrochemicals is also avoided. In conventional practices, disease management is done homogeneously throughout the field. However, the actual disease distribution is heterogeneous because of the microclimate prevalent in the field and diversity in plant-pathogen interactions at the individual level. Consequently, pests and diseases are often manifested in patches, gradients, and random patterns that require precise application of crop protection agents (insecticides, fungicides etc.). Therefore, site-specific crop management is necessary under the precision agriculture approach that aims for the minimum use of agrochemicals. The integration of advanced technologies like Artificial Intelligence (AI) and Machine Learning (ML) to the precision agriculture is hence essential to mitigate yield losses, boost productivity, and fortify our food systems while restricting the use of agrochemicals harmful to the environment.

2. PLANT DISEASES AND THEIR SYMPTOMS

Plant disease refers to the condition when normal growth and development of the plant are hindered and the plant behaves unusually. It can be due to interaction with pathogens or any other abiotic conditions (pH, temperature, nutrient deficiencies etc.) under which the plant has been growing (Savary et al., 2012). Pathogenic infection disrupts the plants' physiological and metabolic balance, ultimately compromising productivity and causing significant yield loss and tissue damage (Martinelli et al., 2015). There are three major disease types of plants - viral, fungal, and bacterial, as caused by the pathogens. These diseases along with the nutrient deficiencies might lead to a variety of visual symptoms. For example, changes in leaf/plant color (yellowing, browning, mosaic) occur due to destruction of pigments (chlorosis) and death of cells/tissues (necrosis, blight). Sometimes, spores cover the surfaces of leaves and other organs (mildews, rust). The morphology and architecture of leaves/plants might change due to viral attack whereas bacteria and fungal pathogens often cause rotting and wilting of plants. At the early stages of disease development, plants might undergo physiological or metabolic alterations which can be detected through advanced laboratory techniques, such as, gene-expression based microarrays, spectroscopy-based methods, volatile organic compound-based analyses that are very costly requiring high-end instruments. For these, continuous and random sampling throughout the field is necessary at regular interval since the sowing of crops, irrespective of disease appearance. However, human eyes can capture the visual symptoms efficiently only at the later stages of disease development when many plants are already affected in the field (Fig. 1). Nevertheless, the advanced imaging sensors like thermal and hyperspectral, can detect the problem well before pathogen gets established and multiplied. Currently, these sensors are mounted on several platforms such as, farm equipment (tractors and other vehicles), air-borne (UAVs or drones), space-borne (satellites) depending on the farm size.

Figure 1. Stages of plant disease development and detection windows by manual and remote technologies (Martinelli et al., 2015; Trippa et al., 2024)

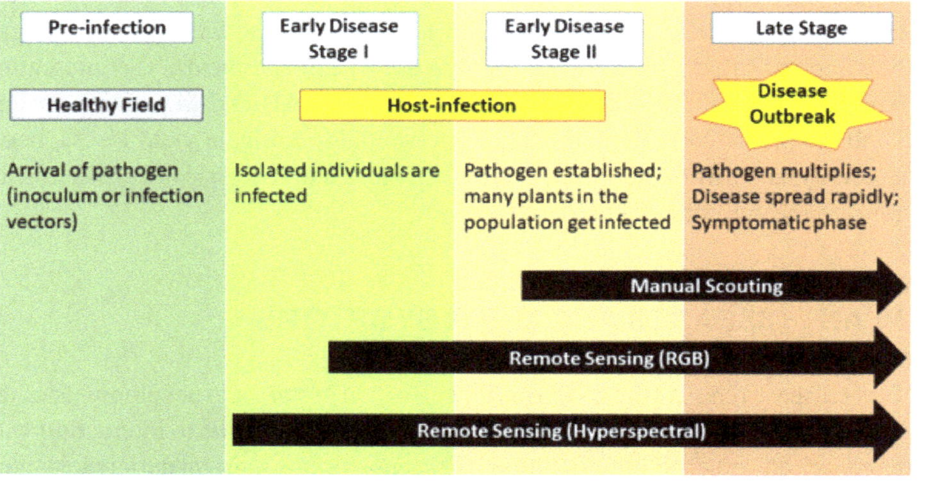

3. REMOTE SENSING TECHNOLOGIES ENABLING AI/ ML APPLICATIONS FOR DISEASE DETECTION

In traditional practices, disease detection was done through manual inspection of the crops in fields by farmers or trained agronomists, followed by laboratory-based diagnostics. These methods are slow, labor-intensive, and usually expensive. Moreover, disease detection through visual observation is highly dependent on the observer's expertise. The delay in detection between initial infection and diagnosis would allow a disease to spread to a larger area. In such cases, AI/ML-based remote sensing technologies have emerged as the alternate, useful options providing timely, accurate, scalable, and affordable solutions for the early detection and management of plant diseases (Trippa et al., 2024). The power of AI/ML models is significantly amplified when integrated with a suite of supplementary technologies that provide the high-quality, multi-modal data needed for comprehensive analysis.

Remote sensing is the acquisition of information about an object or phenomenon by a recording device not in physical or intimate contact with the object or phenomenon under study; hence, it is noninvasive and nondestructive. The technique is suitable for monitoring across various temporal and spatial scales as well as for time-series experiments. Proximal sensing, a sub-area with sensors operating close to the target of interest, is complementary to remote sensing sensu stricto, i.e., monitoring the surface of the earth from satellite and airborne systems. Recent technical advances, especially the introduction of imaging sensors and the development of new methods for data processing (advanced ML, and DL), have considerably increased the potential of remote sensing for plant diseases (West et al., 2019).

3.1 Advanced Imaging Beyond RGB

While standard RGB cameras are the most common data source, other imaging modalities provide richer information. Multispectral and hyperspectral sensors capture image data at specific wavelengths across the electromagnetic spectrum, including those invisible to the human eye (like near-infrared). Healthy and stressed plants reflect light differently. By analyzing the differences in spectral signatures, it's possible to detect physiological changes caused by a disease before visible symptoms like spots or yellowing appear. A review explored the potential of hyperspectral imaging for the pre-symptomatic detection of various plant diseases (Thomas et al., 2018).

3.2 Unmanned Aerial Vehicles (UAVs or Drones) and Satellites

For monitoring large agricultural areas, drones and satellites are useful and are being actively used. Being equipped with RGB, multispectral, or thermal cameras, they can survey hundreds of acres in a single flight. The resulting imagery is stitched together and analyzed by AI/ML algorithms to create detailed "health maps" of the field. These maps can pinpoint disease hotspots, allowing farmers to practice precision agriculture—applying treatments only to affected areas instead of blanket-spraying the entire field. A group of researchers (Abdulridha et al., 2019) applied this approach and used drone imagery to identify avocado trees infected with laurel wilt disease.

3.3 Internet of Things (IoT) and Sensor Networks

IoT networks deploy small, low-cost sensors across a field to continuously monitor variables like air temperature, humidity, soil moisture, and its pH levels. This data, when fed to an AI model, can be used to predict disease risk. For instance, many fungal diseases thrive under specific humidity and temperature conditions. An IoT-based system, like the one developed by Popescu et al. (Popescu et al., 2018) for vineyards, can analyze this data in real-time and issue an alert when conditions are favorable for an outbreak, making preemptive spraying possible and eventually reducing chemical usage.

3.4 Edge AI and Mobile Computing

A major barrier to deployment is the lack of reliable internet connectivity in many rural areas. Edge AI addresses this by performing the AI inference directly on a local device, like a smartphone or a specialized camera, rather than sending data to the cloud. This enables real-time, offline diagnosis. A farmer can simply point their phone's camera at a suspicious leaf, and an on-device model provides an instant analysis.

4. DEVELOPMENT AND ADVANCEMENTS OF AI-ML AND DEEP LEARNING (DL) FOR PLANT DISEASE DETECTION

The application of computational methods in plant pathology started with fundamental image processing techniques, which then led to advanced ML techniques. The typical workflow consists of data acquisition, image preprocessing, feature extraction, model training, prediction and deployment.

4.1 Early Approaches and Foundational Concepts

Before the advent of advanced ML analytics, to automate the identification of visible disease symptoms, researchers developed algorithms that segmented an image (leaf), and then analyzed its properties.

4.1.1 Color and Texture Analysis

Earlier studies, as conducted on rice diseases (Singh & Misra, 2017), focused mainly on extracting color and texture features. The algorithms would convert images to different color spaces (e.g., HSV - Hue, Saturation, Value) to better isolate disease-induced color changes. Texture analysis methods, like Gray-Level Co-occurrence Matrix (GLCM), were used to quantify the roughness, contrast, and homogeneity of a leaf's surface (these features change with fungal or bacterial infections).

4.1.2 Shape and Morphological Features

In other approaches, features such as the area, perimeter, and circularity of spots or blights were measured. They were used as inputs for classification algorithms (Hlaing & Zaw, 2018).

These methods, however, carried certain limitations. For example, they are highly sensitive to variations in lighting, background complexity, and the specific stage of the disease. Such a model developed in a controlled lab environment failed when tested with images from a real field, with its shadows, overlapping leaves, and varying soil backgrounds. Moreover, the process of manually preparing these features was laborious and required deep domain expertise in both botany and computer science.

4.2 The Transition to ML

If datasets are small or handcrafted features (color histograms, texture descriptors like GLCM, shape metrics) are informative, conventional approaches such as Support Vector Machines, K-Nearest Neighbor, Random Forest, and logistic regression are still being implemented. Although they are interpretable and lightweight, their performance is usually poor in comparison to deep networks on big, diverse image sets and require careful feature engineering. Firstly, the ML methods learn the patterns or rules from the training dataset. After that, they apply those patterns/rules to the test dataset. The training process can be achieved in two ways: supervised learning and unsupervised learning.

4.2.1 Support Vector Machines (SVMs)

SVMs are effective in handling high-dimensional feature spaces which make them suitable for such classification problems. A study successfully employed an SVM classifier to identify and categorize diseases on grape leaves based on a combination of color and texture features (Padol & Yadav, 2016). The strength of SVMs lies in their ability to find an optimal hyperplane that separates different classes, making them robust classifiers. However, their performance was still fundamentally capped by the quality of the manually extracted features.

4.2.2 K-Nearest Neighbors (KNN) and Decision Trees (DTs)

This technique, which was used for classification of leaf diseases of paddy plants (Suresha et al., 2017), offers simplicity and interpretability. KNN classifies a new sample based on the "votes" of its nearest neighbors in the feature space. Decision Trees and their ensemble versions, like Random Forest (RF), create a flowchart-like structure to make predictions, which is easy for humans to understand. These models are effective but struggle with the subtle and complex visual patterns inherent in plant disease imagery.

Most works on AI/ML-based plant disease detection used supervised learning such as SVM, DT, RF, Naive Bayes (NB) and multi-layer perceptron (MLP) neural networks (Domingues et al., 2022) (Jackulin & Murugavalli, 2022) (Demilie, 2024) (Subbarayudu & Kubendiran, 2024). All these supervised learning requires manual labeling of diseased areas on the UAV images to train the model. Therefore, the bottleneck with all these ML models remains the same - they are only as good as the data (here features) provided to them. The need for a system that could learn the features directly from the raw pixel data was the primary motivation for the transition to deep learning.

4.3 ML and DL Approaches

The true revolution in image-based plant disease detection arrived with the widespread adoption of deep learning, particularly Convolutional Neural Networks (CNNs). Unlike their predecessors, CNNs do not require manual feature engineering; they learn to identify relevant features—from simple edges and colors to complex lesion patterns—directly from the image data through a process of hierarchical learning (Li et al., 2021).

4.3.1 Convolutional Neural Networks (CNNs)

A CNN is a multi-layered neural network designed to process grid-like data, such as an image. Its architecture is inspired by the organization of the animal visual cortex (Gu et al., 2018). Key components of a CNN include convolutional layers, which apply filters to an image to create feature maps that detect specific patterns (e.g., edges, textures, shapes). Pooling layers then downsample these feature maps, reducing computational complexity and making the detected features more robust to changes in scale and orientation. Finally, fully connected layers act as a classifier, taking the high-level features and outputting a prediction (e.g., the specific disease). The pioneering study (Mohanty et al., 2016) gave a new direction on efficient disease detection. They trained a GoogleNet-based CNN on the public PlantVillage dataset, which contains over 54,000 images. Their model achieved a remarkable 99.35% accuracy in classifying 26 diseases across 14 crop species. This demonstrated conclusively that deep learning could outperform traditional methods by a significant margin. Similarly, Ferentinos (Ferentinos, 2018) evaluated several CNN architectures and found they could diagnose diseases with success rates up to 99.53%, often performing better than human experts under similar conditions.

4.3.2 Advanced Deep Learning Techniques

Building on the success of basic CNNs, researchers have incorporated advanced techniques to improve the performance, especially when dealing with limited or challenging datasets.

- **Transfer Learning:** One of the biggest challenges in agricultural AI is the scarcity of large, labeled datasets for every possible crop and disease combination. Transfer learning provides an elegant solution (Hossen et al., 2025). A model pre-trained on a massive, general-purpose dataset like ImageNet (which contains millions of images across 1,000 categories) already has a well-developed ability to recognize a vast array of visual features. This pre-trained model can then be fine-tuned on a much smaller, specific dataset of plant leaves. Its existing "knowledge" of shapes, textures, and colors provides a powerful head start, dramatically reducing the amount of data and training time required while often improving accuracy. This is the most common and effective approach used in the field today.
- **Vision Transformers (ViTs):** Emerging as a powerful alternative to CNNs, Vision Transformers adapt the transformer architecture, originally designed for natural language processing, to image analysis. Instead of processing pixels with localized filters, a ViT divides an image into a sequence of patches and analyzes the relationships between them using a mechanism called self-attention. This allows the model to capture long-range dependencies across the entire image. Studies like those by Han et al. (Han et al., 2022) have shown that ViTs can achieve performance on par with or even exceeding state-of-the-art CNNs, particularly on large datasets.

While deep learning models offer unparalleled accuracy, they come with trade-offs. They require significant computational resources for training and are often considered as "black boxes," making it difficult to interpret why a model made a particular prediction. This lack of transparency can be a barrier to trust and adoption by farmers. Classical ML models, while less accurate, are generally more interpretable and less computationally demanding.

5. APPLICATIONS OF AI/ML-BASED MONITORING SYSTEMS

The theoretical advancements in AI/ML have translated into a growing number of practical tools that are empowering farmers around the world. These applications range from simple smartphone apps to sophisticated, large-scale agricultural platforms.

5.1 Plantix

Perhaps the most well-known example, Plantix is a free mobile application developed by PEAT GmbH in Germany (Samal et al., 2023). It serves as a "doctor for your plants". A farmer takes a photo of a diseased crop, and the app's AI model analyzes the image to identify the disease or pest. It then provides detailed information about the cause, symptoms, and, crucially, actionable advice on treatment, including both conventional and organic options. The app has a massive user base, particularly in India and other developing nations, and has processed hundreds of millions of images. This creates a powerful feedback loop: every new image submitted helps to improve the AI model, making it more accurate over time.

5.2 Drone-Based Monitoring Services

Companies are now offering drone-based scouting as a service. A licensed pilot flies a drone over a farmer's fields, capturing high-resolution multispectral imagery. This data are uploaded to a cloud platform where AI algorithms analyze it to detect not only diseases but also weed pressure, nutrient deficiencies, and irrigation issues. The farmer receives a detailed report and prescription maps that can be directly loaded into modern farm equipment (like variable-rate sprayers) for precise application of inputs. This approach is gaining significant traction in large-scale commercial farming in North America, Europe, and Australia.

5.3 PlantVillage Nuru - Cassava Disease Detection in Africa

Cassava is a staple food crop for more than 500 million people in Sub-Saharan Africa, but it is highly susceptible to viral diseases like Cassava Mosaic Disease (CMD) and Cassava Brown Streak Disease (CBSD) that can wipe out almost entire harvests (upto 90% yield loss). Traditionally, disease diagnosis was done by the plant pathologists or through lab tests, which were too slow, costly, and not accessible for the majority of rural farmers. Therefore, a real-time, affordable, and scalable solution was necessary. Pennsylvania State University, in partnership with FAO, and International Institute of Tropical Agriculture (IITA), and others, has led the research and innovation project known as PlantVillage and created the AI-powered smartphone app Nuru, which means "light" in Swahili, to identify cassava diseases and to equip farmers with a portable "digital plant doctor" (Mrisho et al., 2020). Researchers, supported by initiatives like Microsoft's AI for Earth, have developed AI models specifically trained to identify CMD and CBSD from leaf images. These models are deployed via smartphones, allowing agricultural extension workers and farmers to quickly diagnose diseases in the field and take immediate action, such as removing infected plants to prevent further spread. This work has been critical in mitigating a major threat to food security in the region.

6. STEPWISE APPROACH FOR A PLANT DISEASE DETECTION MODEL DEVELOPMENT

This section provides detailed steps to prepare a plant disease detection model (Jafar et al., 2024).

Step 1: Problem Definition and Data Acquisition

With an objective to define the scope of the project and gather a high-quality dataset, target crops (e.g., tomatoes, apples, cassava) and their specific diseases are identified. It is started with a limited number to ensure sufficient data is collected for each target disease for model training. Sometimes, public datasets like PlantVillage, PlantDoc, or AI Challenger are utilized for initial prototyping. For a robust model, images are collected in real-field conditions which helps the model in dealing with challenges like complex backgrounds, varying lighting, and different leaf orientations. Proper licensing and permissions are must for all data used. Next, data annotation is done to label each image accurately with the correct disease class. Tools like LabelImg, CVAT, or Roboflow can be used. Consistency in labeling is critical.

Step 2: Data Preprocessing and Augmentation

This step aims to prepare and enhance the dataset to improve model generalization and performance. All images are resized to a fixed dimension (e.g., 224×224, 331×331) as required by the model. Pixel values are then normalized to a standard range (e.g., [0, 1] or [-1, 1]). Data Augmentation is done to expand the dataset artificially to prevent overfitting. Geometric (random rotation, flipping, cropping, shearing) and Photometric (adjusting brightness, contrast, saturation, adding noise) transformations are applied. Sometimes, advanced transformations are made using Generative Adversarial Networks (GANs) like StyleGAN2 to generate high-fidelity synthetic images of rare diseases to balance the dataset. The pre-processed and augmented data is then split into three sets (e.g., 70% training, 15% validation, 15% testing). The test set must be kept completely unseen until the final evaluation.

Step 3: Model Selection

Based on the deployment target (e.g., cloud server, smartphone, embedded device), an appropriate model architecture that balances accuracy, speed, and computational cost is chosen **(Table 1)**. Sometimes, Server/ Cloud Deployment is done for high accuracy. For example, EfficientNetV2 which is excellent as a baseline, deploys a family of models that provide state-of-the-art accuracy and parameter efficiency. On the other hand, Vision Transformers or Swin Transformers can achieve top-tier performance on large datasets by capturing global contextual information. They are computationally expensive but highly accurate. For speed and efficiency Mobile/ Edge Deployment is done. The LDCNet could be an excellent choice with its key innovations noted below:

- Dual-Path CNN and Transformer Design: Extracts both local features (crucial for disease spots) and global dependencies (context of the whole leaf).
- Multi-Scale Feature Learning: Effectively recognizes disease patterns at different scales (small spots, large blotches).
- Lightweight Design: Engineered for low computational cost and high speed, making it ideal for real-time applications on smartphones or IoT devices.
- MobileNetV3 / EfficientNet-Lite: Extremely efficient architectures designed specifically for mobile and edge devices, often used with transfer learning.

Table 1. A comparison of a few ML models with their strengths and limitations.

Model Class	**Example Architectures**	**Best Suited For**	**Strength**	**Limitation**
High-Accuracy	EfficientNetV2, ViT, Swin Transformer	Cloud/Server Deployment	State-of-the-art accuracy	High computational cost, slower inference
Efficient/ Lightweight	**LDCNet**, MobileNetV3, EfficientNet-Lite	Mobile/Edge Deployment	Fast inference, low power consumption	Slight trade-off in accuracy compared to larger models
Legacy (Baseline)	ResNet50, VGG16	Prototyping, and Transfer Learning	Well-understood, extensive resources	Less efficient, often outperformed by newer models

Thus, it is recommended to start with a pre-trained version of a model like EfficientNetV2-S for a strong baseline. For deployment on a phone or drone, then to explore pruning and quantizing this model or switching to a inherently lightweight architecture like LDCNet.

Step 4: Model Training and Optimization

The objective of this step is to train the selected model to learn the mapping from input images to disease classes. In the standard approach, transfer learning is started with a model pre-trained on a massive dataset like ImageNet. This provides a strong feature extraction foundation, allowing the user to achieve high performance with relatively few plant disease images. The loss function uses Categorical Cross-Entropy for multi-class classification. For imbalanced datasets, a weighted cross-entropy loss or focal loss is used to focus learning on harder-to-classify examples. Adam or AdamW are common and effective choices of optimizer. Hyperparameter tuning systematically tunes learning rate, batch size, and data augmentation parameters using the validation set. Tools like Weights & Biases or TensorBoard can help in tracking experiments. Regularization is done by using techniques like dropout, weight decay, and early stopping to prevent overfitting.

Step 5: Model Evaluation

In this step, the model's performance is rigorously assessed on unseen data to ensure it will work in the real world. Key metrics of model evaluation include Accuracy: the overall correctness (can be misleading for imbalanced data); Precision, Recall, and F1-Score (especially important for imbalanced datasets): provide a clearer view of performance per class; Confusion Matrix: reveals which classes are being confused with each other (e.g., is the model confusing two similar-looking diseases?). Finally, Testing performs the final evaluation only on the held-out test set that the model has never seen during training or validation. This gives the best estimate of real-world performance.

Step 6: Interpretation and Explainability (XAI)

This step helps in building trust of end-users (farmers, agronomists) by explaining the model's predictions. Methods like Grad-CAM or Score-CAM is applied to generate heatmaps which highlight the regions of the input image that were most influential in the model's decision. Thus, a farmer can see if the model is correctly focusing on a diseased lesion rather than a dirt spot or shadow, increasing the confidence in the diagnosis.

Step 7: Deployment and Monitoring

In this step, the trained model is integrated into a usable application and maintain its performance over time. The model runs on a server, for example, Cloud API in which users upload images via a mobile app or web interface. Alternatively, on-device (Edge AI) system is available. Here, the model is converted to a format like TensorFlow Lite or ONNX which can run directly on a smartphone. This allows for real-time, offline detection, which is crucial for areas with poor internet connectivity. Then, a user-friendly interface is developed, often in the form of mobile app that allows users to take a picture of a leaf and receive an instant diagnosis and treatment advice. Continuous monitoring is done by logging the predictions and collecting new data. This data is used periodically to retrain the model to improve its accuracy and adapt to new diseases or changing environmental conditions (process known as MLOps).

7. CHALLENGES AND LIMITATIONS OF AI-BASED DISEASE DETECTION

Despite the immense promise and demonstrated success, the path to widespread, equitable adoption of AI-based remote sensing in plant disease detection is fraught with significant challenges. The main issue is lack of high-quality, reliable data, which incapacitate the system. Often the datasets like PlantVillage contain images that are taken primarily in controlled, lab-like settings. There is a scarcity of publicly available datasets captured at large-scale, real-field conditions with complex backgrounds, varying lighting, and multiple diseases on a single leaf. Furthermore, models trained on data from one region often fail to generalize to others because of the differences in crop varieties, pathogen strains, and environmental conditions. Another major hurdle is the lack of interpretability in deep learning models, which is often called the "Black Box" problem. A farmer might receive a diagnosis but has no way of knowing how the AI arrived at that conclusion or solution. This can lead to a lack of trust, especially if the AI's recommendation contradicts traditional farming knowledge. The need for explainable AI where models can point to the features which were used to arrive at the solution or decision can increase transparency and boost trust with the farmers. Deployment and accessibility of the modern AI-based technologies, especially those using cloud-based models, are often hindered by limited internet connectivity, cost of smartphones, drones, and other hardware which are out of bound for the smallholder farmers. Additionally, there is a lack of digital literacy and technical support, making it difficult for farmers to effectively use these tools. Nevertheless, the ethical issue over the ownership of the huge amount of collected farm data is there. Large corporations might exploit the data to their benefits, for instance, in speculating commodity prices. Therefore, ensuring data privacy, security, and equitable benefit-sharing is crucial for building a just and sustainable digital agricultural ecosystem.

8. OPPORTUNITIES AND FUTURE DIRECTIONS

The field is advancing at a rapid pace, with several exciting frontiers that promise to overcome current limitations and unlock new capabilities.

8.1 Generative AI for Data Augmentation

Generative Adversarial Networks (GANs) and other generative models can create realistic, synthetic images of diseased plants. This technology can be used to augment limited datasets, create balanced datasets for rare diseases, and even simulate the appearance of a disease at different stages of its progression, helping to train more robust and accurate detection models.

8.2 Federated Learning for Privacy

To address data privacy concerns, federated learning offers a novel approach. Instead of sending farm data to a central server, the AI model is sent to the local device (e.g., a farmer's phone). The model trains on the local data, and only the updated model parameters (not the raw data) are sent back to be aggregated with updates from other devices. This allows for collaborative model training without compromising the privacy of individual farmers' data.

8.3 Multimodal AI

The future of plant disease detection lies in moving beyond just images. Future systems will integrate data from multiple sources—a concept known as multimodal AI. An advanced model could fuse visual data from a drone, real-time environmental data from IoT sensors, soil health information, and even genomic data from the pathogens themselves to provide a highly accurate, predictive, and holistic assessment of crop health.

8.4 Predictive Analytics and Early Warning Systems

The ultimate goal is not just to detect diseases but to prevent them. By integrating AI models with long-range weather forecasts and remote sensing data, it will be possible to create sophisticated early warning systems that can predict the likelihood of a disease outbreak in a specific region weeks in advance, giving farmers and governments ample time to prepare and implement preventative measures.

The successful integration of AI into agriculture depends on aligning the goals and addressing the concerns of a diverse group of stakeholders. For farmers, particularly smallholders, the primary concerns are practicality, affordability, and trust. A tool must be easy to use, provide tangible economic benefits (either by increasing yield or reducing input costs), and be reliable. They are often skeptical of "black box" technologies and need to see clear evidence of effectiveness before they invest their time and limited resources. The researchers are focused on pushing the boundaries of what is possible. They work on developing new algorithms, improving model accuracy, and publishing their findings in peer-reviewed journals. However, a key challenge for them is bridging the "lab-to-field" gap, ensuring their innovations are relevant and applicable to real-world farming conditions. Currently, the Agri-tech companies and startups are acting as the engines of commercialization by bridging that gap between research and farms. They take the foundational research and build user-friendly products and services, like the Plantix app or drone monitoring platforms. Their focus is on scalability, developing a sustainable business model, and navigating the complex agricultural market while keeping an eye on the changes happening on the fields and in the research sector. Nevertheless, governments and international bodies play a critical role in creating an environment where the other stakeholders can thrive. This includes funding research and development, investing in rural digital infrastructure (like internet connectivity), establishing clear data privacy regulations, and integrating AI tools into national agricultural extension programs to ensure broad and equitable access. The governmental funds to promote agritech companies and startups also support the structure to identify gaps and fill those to support the farming community.

9. CONCLUSION

Artificial Intelligence and Machine Learning are gradually reshaping our approach to plant disease management. The application of AI/ML offers a proactive, data-driven approach for protecting the world's food supply. We have transitioned from basic image processing to complex deep learning models capable of diagnosing diseases with a near immaculate accuracy. When augmented by additional new technologies—from drones and IoT sensors to generative AI—these systems can not just identify diseases but also predict and prevent them, paving the way for a new era of precision agriculture. However, the path forward has its own set of challenges. Issues like data accessibility, model interpretability, digital inclusion, and data ethics must be thoughtfully addressed. The ultimate success of this technological revolution will be assessed not just through academic citations or accuracy benchmarks, but through the tangible impact it can have on the lives of farmers and the global food security. A well-managed, collaborative effort from all stakeholders—researchers building the tools, companies scaling them, policymakers fostering their adoption, and farmers using them to cultivate a healthier and more resilient agricultural future, is required to extract the benefits of these technological advances.

REFERENCES

Abdulridha, J., Ehsani, R., Abd-Elrahman, A., & Ampatzidis, Y. (2019). A remote sensing technique for detecting laurel wilt disease in avocado in the presence of other biotic and abiotic stresses. *Computers and Electronics in Agriculture*, *156*, 549–557. DOI: 10.1016/j.compag.2018.12.018

Demilie, W. B. (2024). Plant disease detection and classification techniques: A comparative study of the performances. *Journal of Big Data*, *11*(1), 5. DOI: 10.1186/s40537-023-00863-9

Domingues, T., Brandão, T., & Ferreira, J. C. (2022). Machine learning for detection and prediction of crop diseases and pests: A comprehensive survey. *Agriculture*, *12*(9), 1350. DOI: 10.3390/agriculture12091350

Ferentinos, K. P. (2018). Deep learning models for plant disease detection and diagnosis. *Computers and Electronics in Agriculture*, *145*, 311–318. DOI: 10.1016/j.compag.2018.01.009

Food and Agriculture Organization of the United Nations. (n.d.). *Agriculture on the proving grounds: Damage and loss*. Retrieved July 24, 2025, from https://www.fao.org/interactive/disasters-in-agriculture/en/

Food and Agriculture Organization of the United Nations. (n.d.). *Plant production and protection*. Retrieved September 1, 2025, from https://www.fao.org/plant-production-protection/about/

Food Security Information Network. (2025). *Global report on food crises 2025*. https://www.fsinplatform.org/report/global-report-food-crises-2025/

Gu, J., Wang, Z., Kuen, J., Ma, L., Shahroudy, A., Shuai, B., Liu, T., Wang, X., Wang, G., Cai, J., & Chen, T. (2018). Recent advances in convolutional neural networks. *Pattern Recognition*, *77*, 354–377. DOI: 10.1016/j.patcog.2017.10.013

Han, K., Wang, Y., Chen, H., Chen, X., Guo, J., Liu, Z., Tang, Y., Xiao, A., Xu, C., Xu, Y., Yang, Z., Zhang, Y., & Tao, D. (2022). A survey on vision transformer. *IEEE Transactions on Pattern Analysis and Machine Intelligence*, *45*(1), 87–110. DOI: 10.1109/TPAMI.2022.3152247 PMID: 35180075

Hlaing, C. S., & Zaw, S. M. (2018). Tomato plant diseases classification using statistical texture feature and color feature. In *2018 IEEE/ACIS 17th International Conference on Computer and Information Science (ICIS)* (pp. 439–444). IEEE. DOI: 10.1109/ICIS.2018.8466483

Hossen, M. I., Awrangjeb, M., Pan, S., & Mamun, A. A. (2025). Transfer learning in agriculture: A review. *Artificial Intelligence Review, 58*(4), 97. DOI: 10.1007/s10462-024-11081-x

Jackulin, C., & Murugavalli, S. J. (2022). A comprehensive review on detection of plant disease using machine learning and deep learning approaches. *Measurement. Sensors, 24*, 100441. DOI: 10.1016/j.measen.2022.100441

Jafar, A., Bibi, N., Naqvi, R. A., Sadeghi-Niaraki, A., & Jeong, D. (2024). Revolutionizing agriculture with artificial intelligence: Plant disease detection methods, applications, and their limitations. *Frontiers in Plant Science, 15*, 1356260. DOI: 10.3389/fpls.2024.1356260 PMID: 38545388

Li, L., Zhang, S., & Wang, B. (2021). Plant disease detection and classification by deep learning—A review. *IEEE Access : Practical Innovations, Open Solutions, 9*, 56683–56698. DOI: 10.1109/ACCESS.2021.3069646

Martinelli, F., Scalenghe, R., Davino, S., Panno, S., Scuderi, G., Ruisi, P., Villa, P., Stroppiana, D., Boschetti, M., Goulart, L. R., Davis, C. E., & Dandekar, A. M. (2015). Advanced methods of plant disease detection: A review. *Agronomy for Sustainable Development, 35*(1), 1–25. DOI: 10.1007/s13593-014-0246-1

Mohanty, S. P., Hughes, D. P., & Salathé, M. (2016). Using deep learning for image-based plant disease detection. *Frontiers in Plant Science, 7*, 1419. DOI: 10.3389/fpls.2016.01419 PMID: 27713752

Mrisho, L. M., Mbilinyi, N. A., Ndalahwa, M., Ramcharan, A. M., Kehs, A. K., McCloskey, P. C., Murithi, H., Hughes, D. P., & Legg, J. P. (2020). Accuracy of a smartphone-based object detection model, PlantVillage Nuru, in identifying the foliar symptoms of the viral diseases of cassava—CMD and CBSD. *Frontiers in Plant Science, 11*, 590889. DOI: 10.3389/fpls.2020.590889 PMID: 33391304

National Aeronautics and Space Administration. (n.d.). *Extreme weather and climate change*. Retrieved September 1, 2025, from https://science.nasa.gov/climate-change/extreme-weather/

Padol, P. B., & Yadav, A. A. (2016). SVM classifier based grape leaf disease detection. In *2016 Conference on Advances in Signal Processing (CASP)* (pp. 175–179). IEEE. DOI: 10.1109/CASP.2016.7746160

Popescu, D., Dragana, C., Stoican, F., Ichim, L., & Stamatescu, G. (2018). A collaborative UAV-WSN network for monitoring large areas. *Sensors (Basel), 18*(12), 4202. DOI: 10.3390/s18124202 PMID: 30513655

Samal, I., Bhoi, T. K., Pradhan, A. K., & Mahanta, D. K. (2023). Plantix app: A success story of artificial intelligence in plant protection. *Van Sangyan*, *10*, 24–44.

Savary, S., Ficke, A., Aubertot, J. N., & Hollier, C. (2012). Crop losses due to diseases and their implications for global food production losses and food security. *Food Security*, *4*(4), 519–537. DOI: 10.1007/s12571-012-0200-5

Singh, V., & Misra, A. K. (2017). Detection of plant leaf diseases using image segmentation and soft computing techniques. *Information Processing in Agriculture*, *4*(1), 41–49. DOI: 10.1016/j.inpa.2016.10.005

Subbarayudu, C., & Kubendiran, M. (2024). A comprehensive survey on machine learning and deep learning techniques for crop disease prediction in smart agriculture. *Nature Environment & Pollution Technology*, *23*(2), 619–632. DOI: 10.46488/NEPT.2024.v23i02.003

Suresha, M., Shreekanth, K. N., & Thirumalesh, B. V. (2017). Recognition of diseases in paddy leaves using KNN classifier. In *2017 2nd International Conference for Convergence in Technology (I2CT)* (pp. 663–666). IEEE. DOI: 10.1109/I2CT.2017.8226213

Thomas, S., Kuska, M. T., Bohnenkamp, D., Brugger, A., Alisaac, E., Wahabzada, M., Behmann, J., & Mahlein, A. K. (2018). Benefits of hyperspectral imaging for plant disease detection and plant protection: A technical perspective. *Journal of Plant Diseases and Protection*, *125*(1), 5–20. DOI: 10.1007/s41348-017-0124-6

Trippa, D., Scalenghe, R., Basso, M. F., Panno, S., Davino, S., Morone, C., Giovino, A., Oufensou, S., Luchi, N., Yousefi, S., & Martinelli, F. (2024). Next-generation methods for early disease detection in crops. *Pest Management Science*, *80*(2), 245–261. DOI: 10.1002/ps.7733 PMID: 37599270

United Nations. (2025). *The sustainable development goals report 2025*. Retrieved September 1, 2025, from https://unstats.un.org/sdgs/report/2025/

United Nations Environment Programme. (n.d.). *How to feed 10 billion people*. Retrieved September 2, 2025, from https://www.unep.org/news-and-stories/story/how-feed-10-billion-people

West, J. S., Bravo, C., Oberti, R., Lemaire, D., Moshou, D., & McCartney, H. A. (2019). The potential of optical canopy measurement for targeted control of field crop diseases. *Annual Review of Phytopathology*, *57*, 267–288. PMID: 12730386

Chapter 6
Leveraging Deep Learning for Real-Time Medicinal Plant Identification

M. Martinaa
http://orcid.org/0000-0003-0977-2106
SASTRA University, India

G. Revathy
http://orcid.org/0000-0002-0691-1687
SASTRA University, India

ABSTRACT

The culture of medicinal plant cultivation has been in existence for many decades among the people of India. Most of the health benefit associated with medicinal plants comes from the Indian Forest. Medicinal plants have, therefore, been the subject of interest to scientists and researchers because of their importance in the survival of man. The identification of medicine plants requires a lot of time and the procedure must be done by experts in the field. Vision-based system can be easily and accurately used to identify the herb plants by scientists as well as the users. This research employs a deep learning approach to design an artificial intelligence system that can be used to identify the herb plants. In this paper, we have selected six plants from Kaggle database that holds maximum benefits which are Indian beech, curry, tulsi, betel, mint, and neem.

DOI: 10.4018/979-8-3373-2429-6.ch006

1. INTRODUCTION

Plants produce oxygen which is a requirement for all living organisms in the earth. Flora comes in all sizes and forms and plays the all-important role of creating oxygen and providing water to all the inhabitants of the earth (Chanyal et al., 2022). In the current study, it was revealed that herbs or medicinal plants have the ability to cure various human ailments (Javid & Haghirosadat, 2017). It is interesting to note that every part of these plants has medicinal properties. Plants have been used by people in many ways in needs such as in cosmetic production, medicine and food preparation. Herbalists apply taxonomy in categorizing the various kinds of medicinal plants that are difficult to distinguish between the other. It is sad though that some nations are still using traditional methods of identifying the herbs by their professional herbalists. Medicinal herbs are those plants that assist both human beings and animals to become healthy. The traditional Chinese medicine used to describe these plants as ordinary plants, however the present generation has named them herbal treatments. Clinicians can administer any part of the plant for preparing the herbal treatment even though this plant type is not popular. The same plant can have several uses in the practical sense of the word. Plants are of immense utility to humanity in that they are used for healing medicine and preparing of food and safe to drink. Man has always looked to nature for cure in the event of one ailment or the other throughout history. Physicians prescribe natural plant-based product to cure medical conditions in patients. Long before the practice of iatrochemistry came to the scene in the 1500s, plants were the only means used to address medical issues (Barimah & Akotia, 2015). Patients are interested in using herbal medicine since they get dangerous side effects of drugs and other treatments do not work. Food and shelter are two resources that are mostly obtained from the plant by most animals in existence. People use coal and natural gas today yet these are products of long-lived plants on which human beings have negatively impacted on plant life and minimized harvest.

This disaster left several severe problems such as environmental destruction and weather shifts that posed risks in the survival and development of people. This is due to the fact that the use of poor-quality chemicals in the preparation of herbal medicine is a real threat to the health of humans and the growth of industries all over the world. Because of this trend, scientists have begun to search for new approaches to categorize the herbal medications. Recent studies show that the plant's leaves enable further research on plants and make plant harvesting easier. This is the basic technique of identification of every medicinal plant. Image processing is a current area of study and development and it is not any longer strange to use computers in plant identification. There are millions of identified plant species in the world and some are poisonous to man while others are used to cure diseases and are facing extinction. Plants maintain the food chain and thus help human beings to survive. According to its uses medicinal plants are used in the following three medical fields; traditional medicine, herbalism, and Ayurveda. Herbal plants are those plants, which have curative properties inherent in the chemical constituents of the plant. The use of traditional medicine in the world is still high at 80% of the population as they depend on it in meeting their health needs. Herbal plants are those whose roots stems or leaves can be used to develop medicine. They should be found in the forests. From observing the leaves they learn many aspects regarding the behavior of the herbs. Researchers have to distinguish plants properly to preserve the plant species (Rao et al., 2022).

Ayurvedic practitioners collected plants for the preparation of medicines for their clients. All these are in vain to date as only a few people practice this tradition today. Ayurvedic medicine annual production has a turnover of more than 4000 crores. Currently, there are 8500 organizations in India that produce Ayurvedic drugs. People have also criticized the materials used in Ayurvedic treatments due to the commercialization of this system of treatment. Currently, women and children go to the natural resource to get the right medicinal herbs since they cannot differentiate the appropriate ones to use. The manufacturing facilities receive deliveries of plants that are not what they agreed with the suppliers on or do not meet their specifications. It is evident that most manufacturing plants do not have the appropriate mechanisms of monitoring the quality of the plant during the process of production. The existence of different names for the same plant in the different regions poses a lot of understanding difficulties. It is then challenging to recognize plants by hand when they come in that condition, that is dried. Incorrect use of medicinal herbs leads to the Ayurvedic treatment procedures not to be effective. This intervention may have side effects that we cannot anticipate and may even lead to some negative consequences. Thus for ayurvedic pharmaceuticals to experience further medical and financial growth the industry must adhere to quality control standards for the medicines and the raw material used in their preparation. It is still challenging and takes considerable time to identify medicinal plants even with the help of experts (Singh & Misra, 2017) because it involves the same procedures as in other plant identification tasks. These challenges led researchers to develop multiple autonomous systems for plant and leaf recognition and they mainly employed machine learning approaches.

2. RELATED WORK

This paper discusses the methodology of identifying plants from the photos of their leaves with the help of current techniques of ML and DL (Chanyal et al.,

2022) and the efficiency of these techniques. This work looks at various images and identify the types of image processing that can be used in the detection of leaves as well as extraction of various measurements of leaves. Diseases are peculiar to certain species of plants and thus their identification early enough is important. Recent advancements in both ML and DL are promising for assisting in identifying and categorizing different symptoms of plant diseases. This analysis looks at all the methods that have been employed in classifying plant leaves using both, ML and DL algorithms.

(Malik, Ismail, Hussein, & Yahya, 2022) According to the study, the author discusses a system that can recognize plants and their local medicinal herbs in real-time. Indeed, the new model of the approach showed better results than the classic model in both datasets by detecting 11% more test cases. The project is unique in the sense that it uses users' feedback and a mobile application to locate plants in Borneo's region. It also revealed its real-time capabilities of detecting medicinal species through the tests conducted earlier. The research team wanted to find out whether the computer systems could identify medicinal plants accurately or not (Valdez et al., 2022). Thus, our work offers a new medicinal herb database with one non-medicinal image class and ten medicinal classes. For the purpose of classifying medicinal plants, we proposed an economical and efficient MobileNetV3 model. The proposed model seems to have better performance in identifying the challenging tasks due to its ability to transfer knowledge from the other tasks. Overall, the findings demonstrate that it is possible and feasible to develop accurate medicinal plant classifier.

This article identifies the ability of convolutional neural networks in recognizing various species of Indian leaves based on the research (Abdollahi, 2022). Currently, the Deep Learning frameworks are implemented in research teams to analyze plant identification and classification. This study established a database of plants that grow wild in the natural world as the main subject of the research. The authors of the research utilized the transfer learning approach by choosing MobileNet as the CNN pre-trained model. That is, our evaluation dataset comprised 3000 photographs of plants belonging to 30 medicinal plant categories. It only took our method to reach 98.05% accuracy in order to pass validation in terms of classifying hidden test data. The author explains one of the methods used in identifying the plant from a single leaf sample (Sivaranjani et al., 2019). ExG-ExR offers better vegetation extraction from photos than conventional vegetation indices at a higher quality. As it was previously mentioned, the OTSU method does not use threshold value since it has a built-in zero threshold. In all lighting conditions, their ExG-ExR index is applicable whereas, using the Otsu method for ExG enhances animal and plant information. It can be seen that our research can discover binary species studies with the help of the ExG-ExR index. The leaves were then divided into individual pictures using a mask tool. The classifying plant species is achieved through input data, which consists of leaf color and texture and by utilising the Logistic Regression model. Based on the results of the current study, it is recommended that a deeper network model is constructed and its settings are optimized for the purpose of improving the results of species recognition (Zin et al., 2020). Therefore, this research establishes that the multi-layer strategy yields superior outcomes when dealing with small samples. They help organizations increase their productivity in all the activities they conduct by applying data augmentation. In case of absence of updates of models with new information, basic transforms such as scaling, adjust orientation, and flipping enhance accuracy. The experiment was given better results with the creation of a new database with plants from Malaysia.

There are six types of medicinal plants that the research team (Saikia et al., 2021) classified by multiple traits of the leaves. First, the system was tested without combining any of the feature attributes where it was able to detect plants and then combined the attributes and improved the results. Out of all the categories of neural networks, the back propagation neural network (BPNN) outperforms all the others in the recognition of leaves. In this experiment, our proposed method yields better outcome with 90 images of leaves belong from six different types. It was possible to identify the differences between the two leaves using the shape, texture, and colour signatures as explained in the following sections. According to the survey discussed in (Sachar & Kumar, 2022), there is an ensemble learning for artificial intelligence's deep learning field that enables users to determine plant species from a captured leaf image. In the given dataset, there are total thirty different types of medicinal plant leaves. First, transfer learning was applied to our neural network model before training. As the relevant database was being passed through a SoftMax-Dense Layer combination, we employed trained deep learning models to extract features from the input images. For this purpose in our tests we employed 3-fold and 5-fold cross validation to get accurate results. They utilized Ensemble DL-Automated Medicinal Leaf Identification (EDL-AMLI) which integrated the result of multiple models. EDL-AMLI outperformed any other advanced pre-trained algorithms in terms of results.

The proposed method in (Manoharan, 2021) aims at increasing the precision of herb detection since the authors of this paper highlight some issues with current datasets. The application of the image segmentation method is more effective when the dimension variables are included in the data sets. The identification of herbal leaves employs the use of an ML classifier and an Exclusive-OR gate to check results when using deep knowledge-based identification. Another method that has been adopted in enhancing identification is the two-stage authentication (TSA) to enhance the identification of the herbal leaves. They can develop a very good and effective system by integrating the machine learning technology with the image segmentation. The most effective methods of extracting leaves enhance the recognition since they filter out leaves from images. The method improves both accuracy and standard performance measures at the same time.

3. PROPOSED METHODOLOGY

In the following sections, we shall identify medicinal herbs in real-time using the Graphic User Interface system. The three phases of research are; training phase, phase for testing, and phase for deployment.

- The data collection in training is from Kaggle website. For the purpose of data organization, we maintain the divisions of training, validation, and testing sets. The information collected is then subjected to data enhancement and resizing within the team. The DL model is trained using the training as well as the validation data. To assess the performance of the model, the research team decides to use four measures such as recall, precision, accuracy and loss.
- After validation and training of the DL model, the team has used the test data to test the model. The DL model is considered ready for production if its output is satisfactory in accordance to the project specifications.
- That is why scientists come across important findings that are still in the stage of preparation for use in practice. One of the key issues of the DL model is that it requires certain level of technical expertise to work on. When the DL model has been developed to the extent of its quality, it is then uploaded to the cloud where streamlit helps to capture real time images to identify medicinal plants. The DL model serves to identify medicinal plant species. The flowchart presented in the figure 1 displays all the previous information and explains how to use Deep Learning for Real-Time Medicinal Plant Identification.

3.1 Data and Its Processing

In our world there are numerous products derived from the plant sources that have medical applications. From time immemorial, medicinal plants have been the form of natural treatment given to humans by doctors. Classification of medicinal plant species helps the pharmaceutical firm to build and sustain its products. The public lacks essential knowledge about how medicinal herbs are identified and utilized. The DL model does address this issue. This is because the training process of a deep learning model is much dependent on the number of pictures of leaves that the model will be fed. The study based its data on Kaggle dataset which was sourced from (Vishnuoum,). Though there is a large number of medicinal plants present on the earth, our concentration will be restricted to six leading species Indian beech curry tulsi mint neem and betel. In the figure, different images are given for each species to illustrate the variations. There are 500 images of each plant species in the Kaggle database. For each species, we collected 350 pictures to divide further into training, validation, and testing sets. The following are the distribution of data details include in this research shown in table 1 and figure 3. The data was divided in a ratio of 70:20:10 for training, validation and testing respectively. The image collection process entails pre-processing of the images that have been collected. Resizing and augmentation belong to the type of activities that are included in the image preparation process. The amount of input images' pixels varies depending on the type of the CNN network (Saponara & Elhanashi, 2022).

The image sizes were reduced to the 224 pixels by 224 pixels because the MobileNet required this format. However, better results in prediction can only be obtained with more training data. We have only 350 images of each species in the collection. Data augmentation is a way to generate new training example from the existing data set. Data augmentation techniques increase the size of the data (Shorten & Khoshgoftaar, 2019). This procedure is used by scientists for most of their tasks while working with small databases. The extra training images provide a barrier against overfitting while the test images with more details give better outcomes. Data augmentation is when the photos are flipped, rotated or have the color altered in order to create more data. The training photos were rotated at an angle of 5, 90 degrees in the one direction and then in the other direction. You can rotate images horizontally or vertically i.e. clockwise or anti-clockwise. Image manipulation is a process, with which we alter an image and in particular change its color. All the manipulation methods including rotations, flips and colour change for all the medicinal plant species used before and after processing are shown in Figure 5.

Table 1. Distribution of medicinal leaf image data

Data	Train	Validate	Test
Indian Beech	350	100	50
Tulsi	350	100	50
Curry	350	100	50
Neem	350	100	50
Betal	350	100	50
Mint	350	100	50

Figure 1. Workflow of Leveraging DL for Real-Time Medicinal Plant Identification

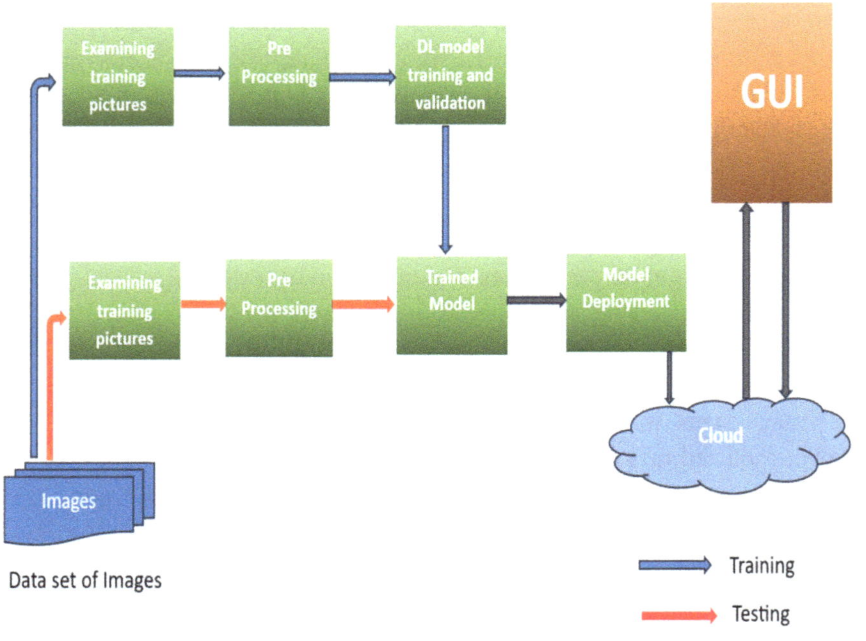

Figure 2. Example of pictures of medicinal leaves

Figure 3. Distribution of data

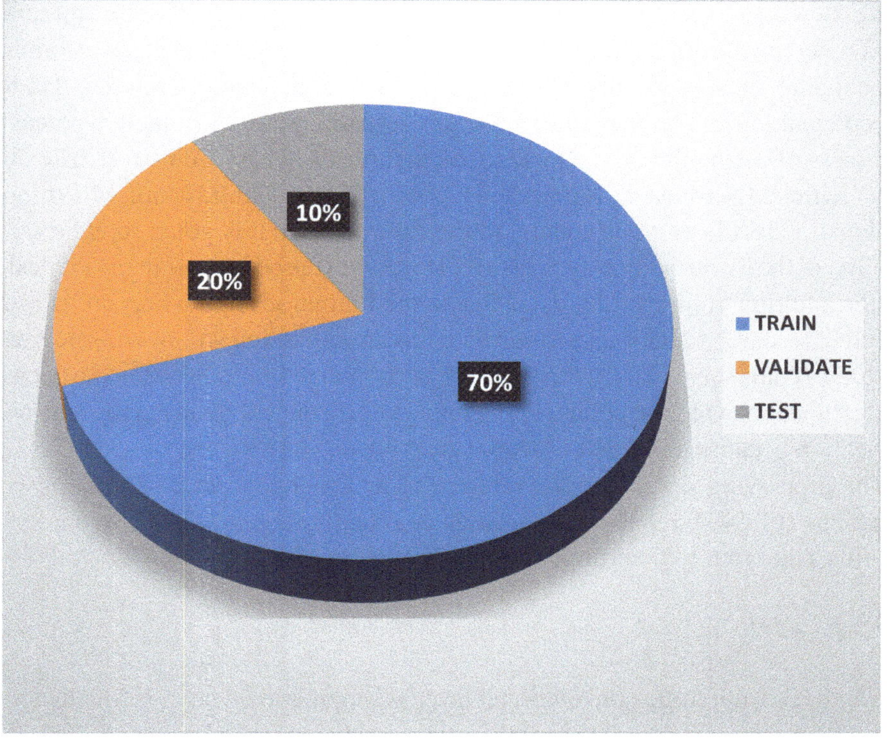

3.2 Building DL Models

(Uma et al., 2022) suggests a new system that uses EfficientB4Net and has a precision of 95% in detecting the medicinal plant. This framework is suitable for real-time use when we couple it with image capturing and processing technologies. The research provides an effective web application in real-time to identify the medical plants with the help of ResNet50v2 and it also gives very less accuracy for 30 species. It is noteworthy that our database is under a constant update concerning the natural environment changes (Goyal et al., 2023). The research team continues to refine its approaches for handling unequal sample sizes and growing the sample database to account for variations in the natural environment. This research study proposed a real-time identification of medicinal plants using Deep Learning EfficientNet-B1 model with 87% accuracy (Malik, Ismail, Hussein, & Yahya, 2022).

(Abirami et al., 2023) focuses on identifying live medicinal plant with the help of EfficientNet and it also elaborates the pros and cons of this new area of research. (Rani et al., 2022) state that real-time plant recognition remains a technical concern as it poses certain technical issues that need to be addressed lacked information on how EfficientNet can be used to recognise plants in real time. Mobile application uses DL techniques of VGG16, InceptionResNetV2 and MobileNet for recognition of medicinal leaves and provide the details. The public cannot browse a database of medicinal leaves and there are only a few research studies available whereas the awareness of the public on the local medicinal plants is very limited. (Kiflie et al., 2024) EfficientNet-based deep learning system for the identification of Ethiopian medicinal plants is quite efficient and the future research is going to enhance the usability of the identification system in real time with the better user interface designs and the addition of more plant data. Due to the limitations of the present dataset in the future research, more data and superior feature extraction will be incorporated in the study. (Gajjar et al., 2022) there may be some merge issues between the datasets; still, Efficient Net has the potential to identify the medicinal plants in the real-world application because of 0.9861 F-score accuracy.

The processing power will be referred to as C. The number of floating-point operations (FLOPs), required for a single convolutional layer can be expressed by equation 1 in order to determine the layer's computational complexity.

$$C=F_s^2 . \text{p} . \text{q} . H_0 \cdot W_0 \tag{1}$$

where, C denotes the convolutional layer's computational cost, F_s Stands for the square convolutional filter's (kernel) spatial dimension (side length). $F_s \, X \, F_s$ is the filter size, A single convolutional filter's weight count is indicated by the F_s^2 value, p Indicates the convolutional layer's input channel count, q Indicates the number of filters or output channels in the convolutional layer, H_0 Indicates the feature map's output height. W_0 Denotes the output feature map's width.

In the convolution process, the input dimensions (H_0 and W_0), the filter size (Fs), the stride (S), and the padding (P) all affect the output feature map's dimensions (H_0 and W_0). For H_0 and W_0 is as follows in equation 2 and 3,

$$H_0 = \left\lfloor \frac{H_i - F_s + 2P}{S} \right\rfloor + 1 \tag{2}$$

$$W_0 = \left\lfloor \frac{W_i - F_s + 2P}{S} \right\rfloor + 1 \tag{3}$$

As suggested by your notation Fs X Fs, if we assume a square output feature map, then

$H_0 = W_0 =$ Fs and the formula becomes simple as follows in equation 4

$$C = F_s^2 . \text{p} . \text{q} . F_m^2 \tag{4}$$

Our model can take images of 224 * 224 with 3 channels which can be in bytes as 224 * 224 * 3 = 150.912 bytes. The input image requires the size of 224 by 224 but the size of this image must be far less than 32. The data from all the three input channels is incorporated according to the third dimension's indication. They are distinguished by the names red, green and blue channels. The proposed starting model has 32 filters that extract features of 3x3 in all the input channels. The basis of EfficientNet is to replace some of the deep convolutional layers with more efficient ones, which is the same as in MobileNets. EfficientNets makes use of Mobile Inverted Bottleneck Convolution (MBConv) blocks that employ depthwise separable convolutions and optimised operations to provide the following results. MBConv blocks usually contain depthwise convolution operation which operates channel-wise independently as the 3x3 convolution. The model then convolves the channels once with a kernel of size 1 and combines elements to form new ones. In order to get better feature information, EfficientNets employ 1×1 convolution layer called expansion layers prior to their depthwise convolutions. EfficientNet is based on a simple concept of using a smaller number of parameters and computational resources as compared to traditional convolutional architectures while maintaining or improving the performance. EfficientNets improve the performance by applying a compound scaling method that scales the depth, width, and resolution of the network.

3.3 The Proposed Methodology

The following can therefore be considered as the general steps in our suggested approach. In the dataset collection phase, our aim is to capture and gather medicinal plants available around Kumbakonam Tamil Nadu, India and some other regions. There are three steps in it. Gathering information is the first step of doing research. taking pictures of plants in their natural environment while taking into account changes in lighting, viewing positions, leaves, flowers, and fruits, as well as backgrounds. Thus, our team will choose ethical means of obtaining plant material for the project. Thus, Herbarium and Internet Resources could be considered as the second step in our approach. It is a study that supplements the data gathered in the herbarium by using digital archives and other online materials available to the public. Seeking for professional help is the next stage of the process. In this way, we will consult with regional specialists in plants and local medicine men in the naming and identification of plant species. Data annotation entails assigning the scientific name of the species as well as any local names of the medicinal plant species that may exist for each image. It is thus important to come up with a comprehensive labelling system that will enable the achievement of accurate and quality data outcomes. Thus, for efficient use of EfficientNet all the images are resized in the same input format. To improve the model's real-world tolerance, apply various image augmentation such as rotating, scaling, cropping, flipping and adjust brightness/contrast to increase the size of the training dataset. It is always beneficial to normalize the pixel values to the range between 0 and 1 so that the training process is faster and the accuracy is higher. Ensure that none of the plant species is over-represented in any of the training, validation, or testing dataset partitions.

Here in figure 4, we have compared different versions of EfficientNet and determine which of them is efficient for real-time application in terms of its accuracy and performance. Consider the options that you have in terms of computers that are available for use which may include cloud servers or mobile devices. The process of training this model utilizes the ImageNet weight data for initializing the EfficientNet architecture. The small size of our medicinal plant dataset helps in training to be complete faster and provides better outcomes. Update specialized layers in the pre-trained network for plant recognition and leave the basic network components unalterable. The EfficientNet model applies the Adam and SGD optimization algorithms together with categorical cross-entropy loss for training. To address this issue of overfitting, the use of weight regularization and early stopping of training when the performance of the validation set starts to degrade should be used. To check if the model is learning properly, compare the results obtained on the training as well as the validation set during the training process.

The optimization phase focuses on how to reduce the model size and the time it takes to make predictions while having high accuracy by comparing between quantization-aware training and post-training quantization. This step minimizes the likelihood of the model's performance being affected by the processing power of the device that it is installed in. To decrease the model size and the power consumption model pruning studies the techniques for eliminating the secondary weight connections from the network. Platforms that enhance model performance depending on specific devices such as mobile GPU and TPUs are discussed in this study including TensorFlow Lite and Core ML. The process of setting up an efficient data pipeline for acquisition of raw real-time images is defined as the 'efficient data flow system'.

3.3 Web-Based DL Model Deployment on Google Cloud Platform

As indicated in the diagram shown in Figure 5, the Web Browser operates as the User Device. It is designed to be web-based and can be run on any device ranging from mobile phones to computers by just using the browser. Users have two options to upload the image of the neem leaf plant, through which the system will analyze it. The web application is the front end of the system that is used by the users. The system we propose has a simple graphical user interface that can be implemented with current HTML/CSS/JavaScript. They Hosted on Google Firebase Hosting or

App Engine. Whenever the user uploads his or her picture it leads to the creation of a Predict button to signify that processing is about to start. The Backend Server is API. For the development of their APIs they used Flask, FastAPI and Node.js. They implemented it via Google Cloud Run and Cloud Functions.

This receives an HTTP request before the uploaded image gets processed through some processes such as resizing and normalization. Thus, the system forwards the image data to the running deep learning model. Google Cloud Storage (GCS). It is used as a temporary storage bin to store uploaded pictures for ease of access by the users or any other use that may be necessary. Our system has the feature of data storage expansion that safeguards the user data efficiently. The proposed system employs a Vertex AI to host either the pre-trained or customer's trained EfficientNet model. It takes input from its source in a processed image.

Thus, depending on the Neem leaf data, the system identifies which plant type is present in the sample. It sends back prediction results. Firestore is employed as the Cloud Database service in our solution for the application. It stores information about users and their plant submissions, plant details, and such other things as the medicinal values of Neem. This system allows the identification of a plant and the retrieval of its data at a very short time. The web app receives data from the next step which is aimed at sending back the prediction results. It is depicted as the plant name Neem, and its uses in health such as it has antibacterial properties and skincare among others. The system has a pop-up window or card that offers the additional herbal information when the user hovers the mouse over an image of the plant.

Google Cloud Platform is an effective platform for implementing both EfficientNet and other DL models for use in web applications. Due to GCP's scalability, it is ideal for handling data preparation processes required to bring about these models. GCP provides reliable means to save the trained EfficientNet models and deploy them for processing and analyze the results using the secure cloud services. Google Cloud Storage is also able to offer a cheap solution to store EfficientNet model artifacts in a global level. This is because the web application established with the users allows them to access the required model files from any location securely. Cloud storage is useful in management of security of data that is processed by our web application by providing secure storage.

Figure 4. Web-Based DL Model Deployment on Google Cloud Platform

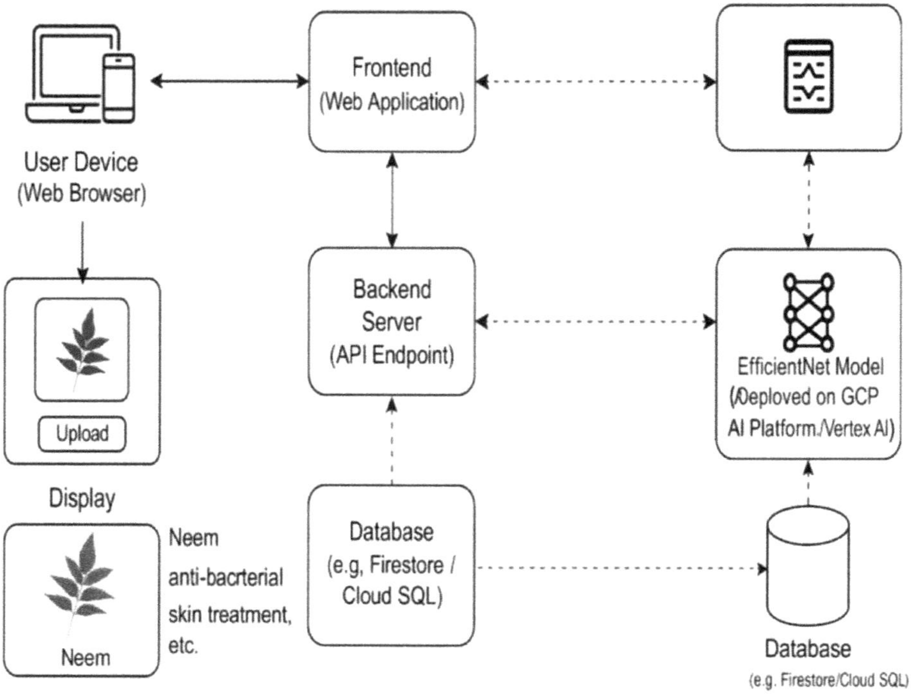

GCP provides various services which are useful to develop our web application along with integrating EfficientNet model. We use Google Cloud Platform tools such as Compute Engine, Kubernetes Engine, Cloud Run, and Cloud Functions to host the application's logic that processes user input before passing them to the EfficientNet model for processing and to return the results. Deep learning systems provide high accuracy but at the same time, they are costly and may not be suitable for a web application that requires them to handle numerous requests from the users. The EfficientNet models have better accuracy in comparison to regular deep networks while consuming fewer parameters and less computational power. Less parameters prevent us from using models that require special hardware that is expensive to own and limits the places where we can place our models. To build an efficient web-based model in GCP, we are required to use existing platforms for deployment and improving the inference system. Converting our EfficientNet model which has been trained into TensorFlow Lite will assist in enabling it to be used in our GCP web application backend. The goal of our work will be to create a web-based DL application with the help of Efficientnet while minimizing the consumption that the application requires in terms of resources for maximum performance.

Figure 5. Before and after augmentation of medicinal visuals

Figure 6. Architecture diagram of EfficientNet

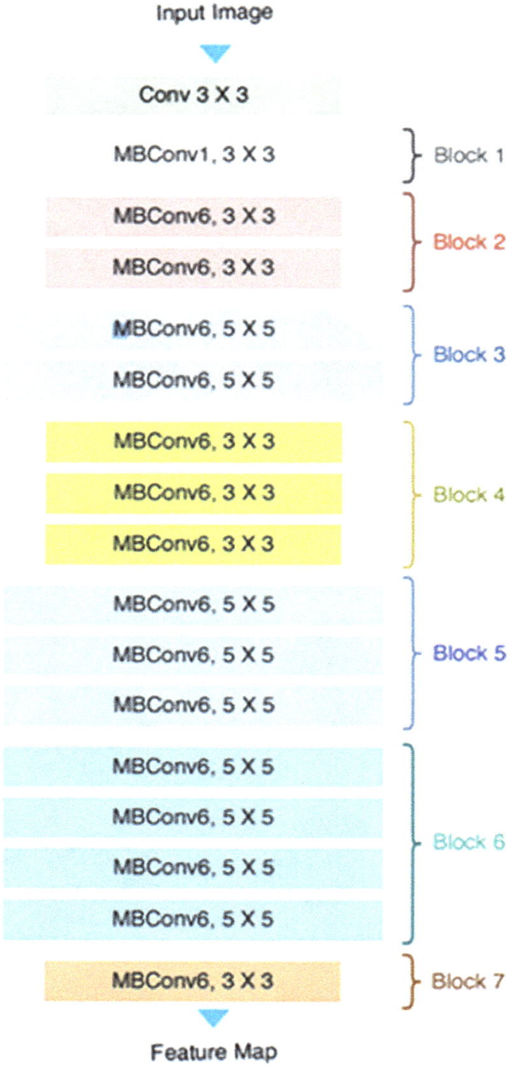

4. RESULTS AND DISCUSSION

In this research work, the goal is to identify the medicine plants in real time. To achieve this task, we used all the required medicinal leaf pictures which are found in Kaggle dataset. These images were captured in different sizes from the original source hence the need to resize them to the same size as shown in the figure below. Since the model was not able to predict correctly the geometrical changes needed in the real world, to the images of the leaves, more data had to be fed to the model. The DL model is trained to 15 epochs using the augmented plant leaf images.

4.1 Training and Validation Result

In the case of a DL model, there are loss, recall, accuracy, and precision to evaluate the outcomes. All the measured metrics results are given in Table 3. The training loss rates of the DL model is at 1.01 - 0.08 while its validation loss rates is 1.58 - 0.16 from the 0th to the 15th epoch. The recall for training is at 0.55 at epoch 0 and rises to 0.98 at the final epoch while validation recall is at 0.69 at epoch 0 and 0.96 at the final epoch. At the same time, in the third stage, we identify the performance of the deep learning model. The DL model proves an accuracy of 0.65 during the training and validation of the epoch 0. From the above model, it can be observed that the training accuracy and validation accuracy is at the 14th epoch with 0.98 and 0.96 respectively. Once training and validation processes are over the measurement of the model is made. In training phase, the precision is at 0.82 in the beginning and at the end it is 0.98 but in validation the precision is 0.72 at start and ends at 0.97.

Figure 7 below represents the line charts derived from the values presented in the Figure 8. The EfficienctNet loss values are presented in the Figure 7 while the recall values are presented in Figure 9. The performance of the system is depicted in figures 10 and 11. The trend depicted in all the graphs is that of an increase in the given metrics with an increase in epoch numbers across the plot. All the other values but the loss value increase with the growth of the number of training cycles.

Figure 7. Web-Based DL Model Performance Loss

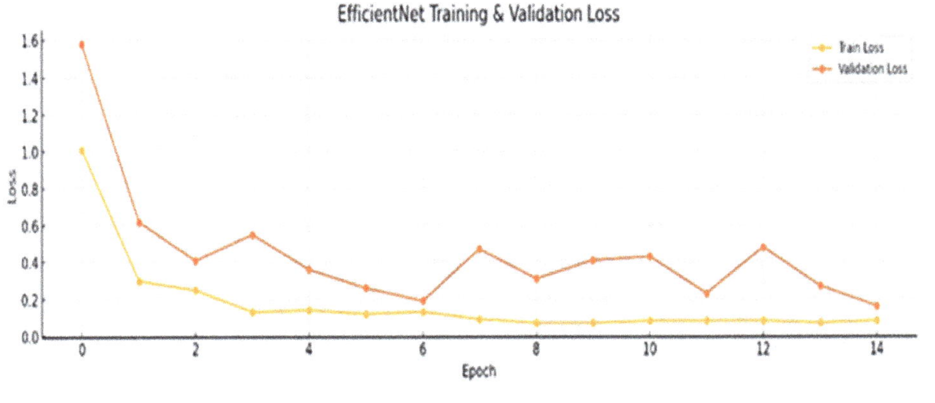

Figure 8. Web-Based DL Model Result

Epoch	LOSS		RECALL		ACCURACY		PRECISION	
	Train	Validate	Train	Validate	Train	Validate	Train	Validate
0	1.0136	1.5813	0.5579	0.6911	0.6525	0.7079	0.8234	0.7288
1	0.32	0.6249	0.8994	0.8233	0.9043	0.8337	0.9265	0.8568
2	0.2539	0.4153	0.9196	0.8682	0.928	0.8778	0.9374	0.8919
3	0.14	0.553	0.9615	0.9451	0.9659	0.9523	0.9693	0.9611
4	0.1454	0.3693	0.9599	0.8461	0.9646	0.8549	0.96	0.8877
5	0.1243	0.26721	0.9645	0.8966	0.9688	0.8918	0.9719	0.9082
6	0.1328	0.1918	0.9636	0.9579	0.9678	0.9699	0.9731	0.9636
7	0.0944	0.4725	0.978	0.8285	0.9804	0.8349	0.9821	0.8415
8	0.0739	0.3184	0.9812	0.913	0.9834	0.921	0.9849	0.9346
9	0.0797	0.4177	0.9851	0.9014	0.9878	0.909	0.9894	0.918
10	0.0844	0.438	0.9816	0.8878	0.9837	0.883	0.9863	0.8914
11	0.0871	0.2324	0.9812	0.9451	0.9831	0.9475	0.9848	0.9537
12	0.0855	0.4826	0.9824	0.8974	0.9849	0.8938	0.9863	0.9081
13	0.0702	0.2709	0.9836	0.9387	0.9847	0.9315	0.9866	0.9468
14	0.0851	0.1619	0.9832	0.9667	0.9843	0.9623	0.986	0.9757

Figure 9. Web-Based DL Model Performance Recall

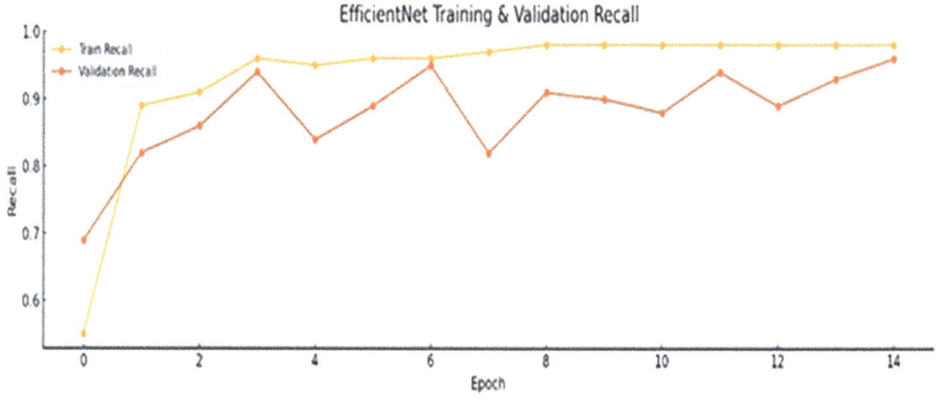

Figure 10. Web-Based DL Model Performance Accuracy

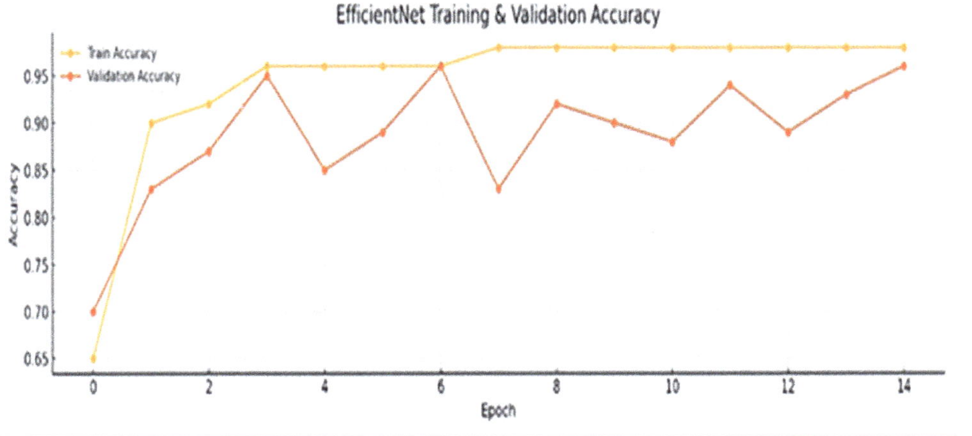

Figure 11. Web-Based DL Model Performance Precision

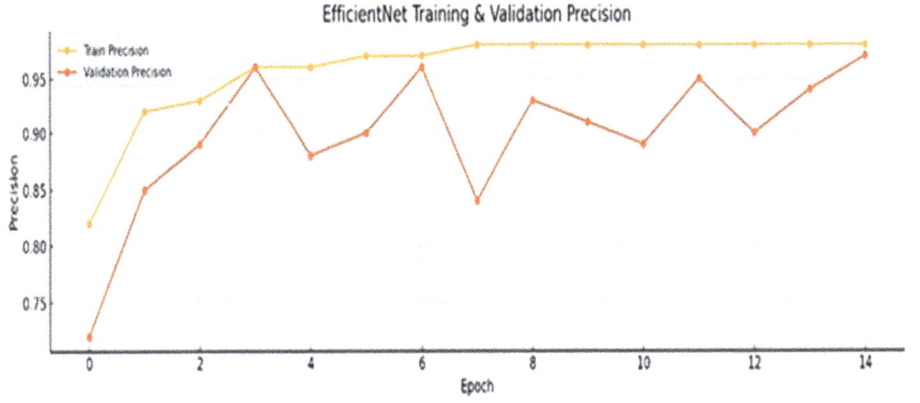

4.2 Result of Testing Phase

After validating our model, we make the test with the data of the test set. In the process of test analysis, we use three parameters: accuracy, precision and recall. The results of the test show that our DL model has an accuracy of 96%, the recall is 96% while the precision is 97%. Therefore, it can be concluded that the EfficientNet successfully recognizes the medicinal plants as it received an average accuracy of over 96.33% on all the metrics used in the experiment. Figure 13 and Figure 12 also presents the bar graph of the DL model test results.

Figure 12. Performance of EfficientNet on Test data

Model	Accuracy	Recall	Precision
EfficientNet	96%	96%	97%

Figure 13. Performance of EfficientNet on Test data

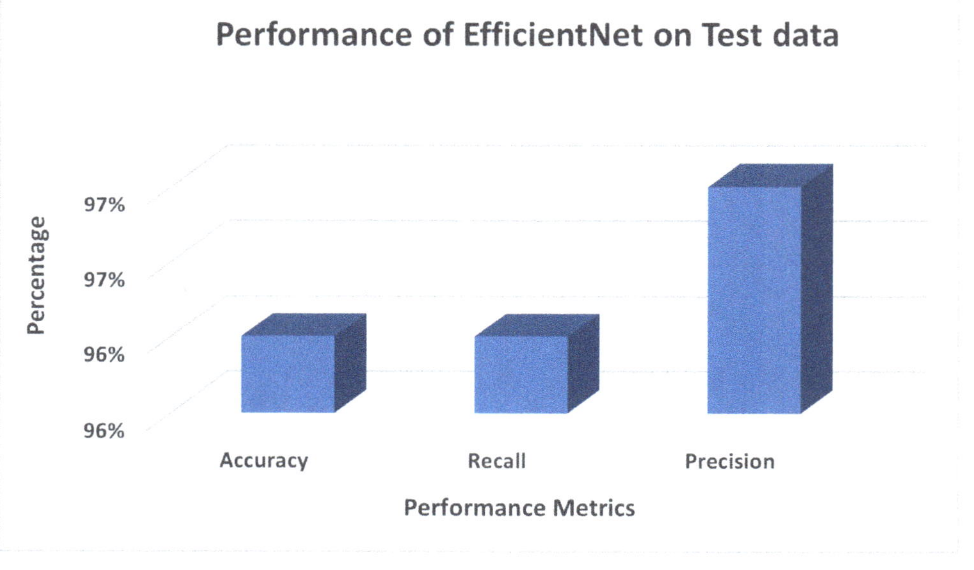

4.3 Deployment

The testing of our built model was proved to be satisfactory as demonstrated in figure 14 during the testing phase. Our team considered and chose the Efficient-Net deep learning model to deploy on Google Cloud Platform for web access on a large scale. We started developing its web application that would fulfil the needs of targeted users. The simplicity of controls of the system was a basic design that enhanced its functionality of the system. When users are in the web app, they are presented with a form to either login or register, the registration form involves a few fields. The users after logging in are directed to the main page of the application where they can only 'scan' or 'upload' an image of a leaf. The application allows the users to capture a picture of a leaf using the camera of their device or use a picture that has already been saved in the device. When a user provides an image to the system, the prediction button becomes clickable so that the user can feed the image through the EfficientNet model. The deep learning model reads the image to get the unique name of the healing herb or plant. Finally, the output is passed to the web application after the processing is complete is done by the cloud. The application also displays the name of the plant as well as, some of the health benefits once it is processed. With us, the web system was tested using images of the tulsi plant. The system was able to identify the plant and a small window appeared indicating the uses of the plant in medicine and benefits that can be enjoyed, through the cloud interface connected to the internet.

Figure 14. Working of web-based web application

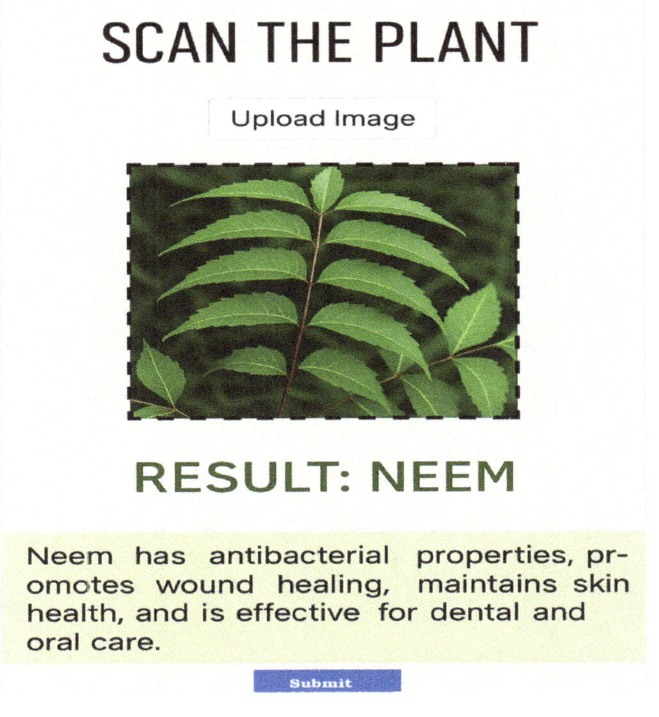

5. CONCLUSION

In as much as life exists, plants are very crucial in the survival of human beings. Indigenous populations for many generations have depended on the use of herbs from their experiences. Herbs are recognized by the healthcare professionals through their experience of taste and smell. With the help of the contemporary techniques there is easy identification of the different herbs in any given sample. In fact, many of the community members including the newcomers consider this method very helpful. The laboratory-based analysis entails sample collection and data interpretation skills which take more time than normal operations. There is therefore the need to develop a fast and correct method of recognizing herbs. This work aims at designing an automatic medicinal plant detector using deep learning techniques. This research project focuses only on six herbs only. It is the new DL system that is capable of producing identification outcome with a reliability of 96.33%. This model will lead to the development of a web-based tool that can identify medicinal plants at a fast rate and the tool will be hosted on cloud services. The application of mobile DL allows the identification of medicinal plants in a shorter time than by using the measuring instruments. Further research will improve the identification skills of the model by including more species of plants.

REFERENCES

Abdollahi, J. (2022). Identification of medicinal plants in Ardabil using deep learning. In *2022 27th International Computer Conference, Computer Society of Iran (CSICC)* (pp. 1–6). IEEE.

Abirami, A., Bhuvaneswari, S., Aadhithya, R., Devesh, R., & Deexith, P. (2023). *An empirical analysis on medicinal plants associated with impact on biodiversity, types, challenges, and technologies.*, DOI: 10.1109/ICACRS58579.2023.10404828

Barimah, K. B., & Akotia, C. S. (2015). The promotion of traditional medicine as enactment of community psychology in Ghana. *Journal of Community Psychology, 43*(1), 99–106. DOI: 10.1002/jcop.21687

Chanyal, H., Yadav, R. K., & Saini, D. K. J. (2022). Classification of medicinal plants leaves using deep learning technique: A review. *International Journal of Intelligent Systems and Applications in Engineering, 10*(4), 78–87.

Gajjar, V. K., Nambisan, A. K., & Kosbar, K. L. (2022). Plant identification in a combined-imbalanced leaf dataset. *IEEE Access : Practical Innovations, Open Solutions, 10*, 37882–37891.

. Goyal, H., Garg, B., & Goyal, A. K. (2023). HerbaVisionNet: Optimized CNN and ResNet50V2 model for enhanced medicinal plant identification and application design. *International Journal of Membrane Science and Technology*. DOI: 10.15379/ijmst.v10i3.3702

Javid, A., & Haghirosadat, B. F. (2017). A review of medicinal plants effective in the treatment or apoptosis of cancer cells. *Cancer Press Journal, 3*(1), 22–26. DOI: 10.15562/tcp.41

Kiflie, M., Sharma, D., Haile, M. A., & Srinivasagan, R. (2024). EfficientNet ensemble learning: Identifying Ethiopian medicinal plant species and traditional uses by integrating modern technology with ethnobotanical wisdom. *Computers, 13*(2), 38. DOI: 10.3390/computers13020038

Malik, O. A., Ismail, N., Hussein, B. R., & Yahya, U. (2022). Automated realtime identification of medicinal plants species in natural environment using deep learning models—A case study from Borneo Region. *Plants, 11*(15), 1952. DOI: 10.3390/plants11151952 PMID: 35956431

Malik, O. A., Ismail, N., Hussein, B. R., & Yahya, U. (2022). Automated real-time identification of medicinal plants species in natural environment using deep learning models—A case study from Borneo Region. *Plants*, *11*(15), 1952. DOI: 10.3390/plants11151952 PMID: 35956431

Manoharan, J. S. (2021). Flawless detection of herbal plant leaf by machine learning classifier through two stage authentication procedure. *Journal of Artificial Intelligence and Capsule Networks*, *3*(2), 125–139. DOI: 10.36548/jaicn.2021.2.005

Rani, L., Devika, G., Karegowda, A. G., Vidya, S., & Bhat, S. (2022, April). Identification of medicinal leaves using state of art deep learning techniques. In *2022 IEEE International Conference on Distributed Computing and Electrical Circuits and Electronics (ICDCECE)* (pp. 1-5). IEEE.[

Rao, R. U., Lahari, M. S., Sri, K. P., Srujana, K. Y., & Yaswanth, D. (2022). Identification of medicinal plants using deep learning. *International Journal for Research in Applied Science and Engineering Technology*, *10*(4), 306–322. DOI: 10.22214/ijraset.2022.41190

Sachar, S., & Kumar, A. (2022). Deep ensemble learning for automatic medicinal leaf identification. *International Journal of Information Technology : an Official Journal of Bharati Vidyapeeth's Institute of Computer Applications and Management*, *14*(6), 3089–3097. DOI: 10.1007/s41870-022-01055-z PMID: 35975198

Saikia, A. P., Hmangaihzuala, P. V. L., Datta, S., Gope, S., Deb, S., & Singh, K. R. (2021). Medicinal plant species classification using neural network classifier. In *2021 6th International Conference on Communication and Electronics Systems (ICCES)* (pp. 1805–1811). IEEE. DOI: 10.1109/ICCES51350.2021.9488971

Saponara, S., & Elhanashi, A. (2022). Impact of image resizing on deep learning detectors for training time and model performance. In *Applications in Electronics Pervading Industry, Environment and Society: APPLEPIES 2021* (pp. 10–17). Springer International Publishing. DOI: 10.1007/978-3-030-95498-7_2

Shorten, C., & Khoshgoftaar, T. M. (2019). A survey on image data augmentation for deep learning. *Journal of Big Data*, *6*(1), 1–48. DOI: 10.1186/s40537-019-0197-0

Singh, V., & Misra, A. K. (2017). Detection of plant leaf diseases using image segmentation and soft computing techniques. *Information Processing in Agriculture*, *4*(1), 41–49. DOI: 10.1016/j.inpa.2016.10.005

Sivaranjani, C., Kalinathan, L., Amutha, R., Kathavarayan, R. S., & Kumar, K. J. J. (2019). Real-time identification of medicinal plants using machine learning techniques. In *2019 International Conference on Computational Intelligence in Data Science (ICCIDS)* (pp. 1–4). IEEE. DOI: 10.1109/ICCIDS.2019.8862126

Uma, K. V., Bama, B. S., Sabarinathan, D., & Roomi, S. M. M. (2022). Identification and retrieval of medicinal plants of Southern India using EfficientB4Net. *Journal of Intelligent & Fuzzy Systems*, *42*(6), 5097–5112. DOI: 10.3233/JIFS-211426

Valdez, D. B., Aliac, C. J. G., & Feliscuzo, L. S. (2022). Medicinal plant classification using convolutional neural network and transfer learning. In *2022 IEEE International Conference on Artificial Intelligence in Engineering and Technology (IICAIET)* (pp. 1–6). IEEE. DOI: 10.1109/IICAIET55139.2022.9936868

Vishnuoum. (n.d.). *Medicinal plant dataset (augmented)* [Data set]. Kaggle. https://www.kaggle.com/datasets/vishnuoum/medicinal-plantdataset-augmented

Zin, I. A. M. D., Ibrahim, Z., Isa, D., Aliman, S., Sabri, N., & Mangshor, N. N. A. (2020). Herbal plant recognition using deep convolutional neural network. *Bulletin of Electrical Engineering and Informatics*, *9*(5), 2198–2205. DOI: 10.11591/eei.v9i5.2250

Chapter 7
IoT-Based Fertilizer Pouring

Rajaa Vikhram Yohanandhan
http://orcid.org/0000-0001-6786-3378
SRM Institute of Science and Technology, India

P. Deepak
SRM Institute of Science and Technology, India

A. Ashik Ahamed
SRM Institute of Science and Technology, India

S. Vimal
SRM Institute of Science and Technology, India

ABSTRACT

This report presents the design and implementation of an Internet of Things (IoT) system for automated fertilizer pouring in agricultural applications. The system leverages sensors, actuators, and cloud connectivity to optimize fertilizer application, enhance crop yield, and minimize environmental impact. Precisely delivers the required amount of fertilizer based on pre-programmed schedules or real-time sensor data. Enables farmers to remotely monitor system status, adjust settings, and receive alerts via a mobile app or web interface. Collects and analyzes sensor data to provide valuable insights into crop health, optimize fertilization strategies, and improve overall farm efficiency .Its excellent battery life ensures efficient operation without extended downtimes for recharging, maintaining the effectiveness of the spraying process. By offering an efficient solution, this agricultural robot

DOI: 10.4018/979-8-3373-2429-6.ch007

represents a significant advancement in integrating modern technologies into farming, addressing operational challenges, and potentially transforming traditional agricultural practices.

1. INTRODUCTION

The agricultural sector is undergoing a significant transformation, driven by the convergence of cutting-edge technologies (Niroui et al., 2019). Among these, the Internet of Things (IoT) and electric vehicles (EVs) are poised to revolutionize farming practices, offering unprecedented levels of efficiency, sustainability, and precision. Traditional agricultural methods often involve manual labor, leading to inefficiencies, inconsistent results, and potential environmental harm (Mehra, 2020). The overuse of fertilizers, for instance, can contaminate water sources and degrade soil quality. Moreover, the reliance on fossil fuel-powered machinery contributes to greenhouse gas emissions, exacerbating climate change (Martin et al., 2022). It operates autonomously along designated crop paths, requiring human intervention only at the initial setup, which significantly reduces labor costs and dependence on manual operations. The integration of IoT and EVs presents a powerful solution to these challenges (Murphy, 2019). IoT enables real-time monitoring of various agricultural parameters, such as soil moisture, nutrient levels, and weather conditions. This data-driven approach allows farmers to make informed decisions about fertilizer application, optimizing yields while minimizing waste and environmental impactBy precisely applying fertilizers and pesticides, the robot minimizes waste and reduces harmful leaching, protecting water sources and promoting sustainable practices in line with precision agriculture (Kootstra et al., 2020)-(Brooks, 1986). Its robust battery life allows for extended operation without frequent recharging, ensuring continuous and efficient performance. Designed to be both affordable and easy to use, this agricultural robot offers an accessible solution to farmers, addressing common challenges of high costs, complexity, and slow adoption of advanced technologies in the sector (Kaelbling et al., 1987)-(Gat, 1998).

1.1 Project Description

This module is responsible for assessing the environmental conditions to determine whether they're suitable for toxin operation. The system uses temperature and moisture detectors to cover the girding atmosphere. By collecting real- time data on these environmental factors, the robot can make informed opinions about when and where to spot the toxin. For case, the system may be programmed to avoid scattering when moisture situations are too high or when temperatures are outside the optimal range for toxin effectiveness. This module ensures that the robot only applies diseases when conditions are ideal, promoting more effective use of coffers and precluding damage to crops or the terrain.

2. NAVIGATION AND MOVEMENT CONTROL MODULE

This module is responsible for the robot's movement within the agrarian field. The robot can be controlled through two modes independent navigation or homemade control via a pall garçon using the Internet of effects(IoT). The Arduino Uno serves as the primary regulator, recycling inputs from detectors, motors, and other tackle factors to control the robot's movement. It receives commands from the ESP8266, which handles the communication between the robot and the pall garçon. The robot can followpre-programmed paths or acclimate its route grounded on real- time environmental conditions or stoner input, making it adaptable to colorful field layouts and conditions. Navigation and Movement Control Modules are essential for the development of autonomous systems. By integrating sensor data, planning paths, and controlling movement, NMCs enable vehicles and robots to navigate and operate independently in complex environments

2.1 Handicap Discovery and Avoidance Module

Equipped with detectors similar as ultrasonic or infrared detectors, this module allows the robot to descry obstacles in its path and avoid collisions. As the robot autonomously navigates the field, it continuously scans the terrain for walls similar as jewels, trees, or other obstacles that could stymie its movement. Once an handicap is detected, the robot adjusts its course to bypass it, icing smooth operation and reducing the threat of damage to the robot or girding crops. This module ensures that the robot can operate efficiently in dynamic field surroundings without taking homemade intervention to clear obstacles, enhancing both the robot's trustability and functional safety. By implementing a robust handicap discovery and avoidance module, organizations can ensure a high-performing and resilient network that meets the demands of modern businesses. Handicap Discovery and Avoidance Module: A Detailed Over.Handicap discovery and avoidance is a crucial aspect of network management, especially in large and complex networks. It involves identifying network devices with suboptimal configurations or performance issues that can hinder overall network efficiency. Network Monitoring: Continuous monitoring of network devices, traffic patterns, and performance metrics. Performance Analysis: Analyzing collected data to identify devices with abnormal behavior, such as high latency, packet loss, or CPU utilization.

2.2 Block Diagram Working

In this design, an independent toxin sprayer for husbandry is developed using an Arduino Uno as the main regulator and an ESP8266 as the slave regulator. The system's primary thing is to automate the process of toxin operation grounded on environmental conditions. The robot integrates temperature and moisture detectors to assess the terrain's felicity for toxin spraying. When the environmental conditions are optimal, the robot autonomously navigates the agrarian field and applies toxin precisely where demanded. The robot's movement is controlled either through a pall garçon via the Internet of effects(IoT) or autonomously, depending on stoner preference. Also, the robot is equipped with handicap discovery capabilities to navigate around walls on the land, icing smooth operation without collisions. The use of the Arduino Uno allows for effective processing and control of the robot's functions, while the ESP8266 facilitates communication between the robot and the pall .

3. HARDWARE DESCRIPTION:12V BATTERY

A 12V battery refers to a type of battery that provides a nominal voltage of 12 volts. These batteries are commonly used in a variety of applications, including: Automotive applications 12V lead-acid batteries are widely used in vehicles to power the electrical system, including the starter motor, lights, and other accessories .Solar energy systems 12V batteries are often used in off-grid solar power systems to store energy for later use .Electric bikes and scooters Many electric vehicles like bikes, scooters, and even small electric cars use 12V batteries .Power backup systems 12V batteries can be part of an uninterruptible power supply (UPS) to provide backup power in case of an outage.

Arduino is an open-source electronics platform that makes it easy for anyone to create interactive electronic projects. Whether you're a beginner or an experienced engineer, Arduino allows you to build simple or complex systems using easily accessible hardware and software. At its core, Arduino consists of a microcontroller—a small, programmable computer on a single chip—that can be used to control various electronic components like lights, motors, sensors, and displays. The platform provides a simple way to interact with the physical world, allowing you to create everything from basic gadgets to advanced systems like robots or home automation systems.

3.1 Arduino Uno

The Arduino Uno is a highly accessible, flexible, and powerful platform for anyone interested in electronics and programming.

Figure 1. Arduino Uno is a microcontroller board grounded on the Atmega328P

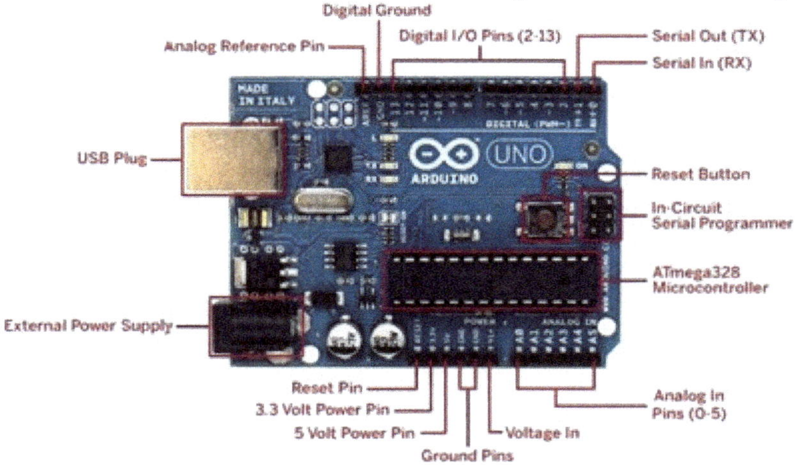

Whether you're a hobbyist, student, or professional, the Uno provides a solid foundation for learning, prototyping, and creating innovative projects as shown in fig.3.1. Its ease of use, open-source nature, and wide community support make it an ideal choice for building both simple and complex projects. Open-Source Hardware and Software: Both the hardware (the Arduino boards) and the software (Arduino IDE) are open-source, meaning you can freely modify and share designs, code, and projects Easy to Use: Arduino's user-friendly software (IDE) allows you to write, compile, and upload code to the board with just a few clicks. This simplicity makes it accessible for beginners, while its versatility appeals to experienced users as well.

3.2 Hardware

The atmega328p board is a well-designed micro controller-based improvement board that provides a variety of components and interfaces for building electronic projects. Below is an overview of the key hardware components found on the Arduino Uno board

1. Microcontroller: ATmega328P

Description: The ATmega328P is the central microcontroller on the Arduino Uno, responsible for executing your program (sketch) and interacting with the board's I/O pins.

Features: It runs at a16 MH clock speed and comes with:

32KB flash memory (for storing the program).

2KB SRAM (for storing runtime data).

1KB EEPROM (for storing persistent data).

Table 1. Technical Specifications

Microcontroller	**Atmega328P**
Operating Voltage	5V
Input Voltage (recommended)	7-12V
Input Voltage (limit)	6-20V
Digital I/O Pins	14 (of which 6 provide PWM output)
PWM Digital I/O Pins	6
Analog Input Pins	6
DC Current per I/O Pin	20 Ma
DC Current for 3.3V Pin	50 Ma
Flash Memory	32 KB (Atmega328P) of which 0.5 KB used by boot loader
SRAM	2 KB (Atmega328P)
EEPROM	1 KB (Atmega328P)
Clock Speed	16 MHz
LED_BUILTIN	13
Length	68.6 mm
Width	53.4 mm
Weight	25 g

3.3 Programming

The Arduino Uno can be programmed with the(Arduino Software(IDE)). elect" Arduino/Genuino Uno from the Tools> Board menu(according to the microcontroller on your board). For details, see the reference and tutorials. Once you're comfortable with basic programming, you can try more advanced projects by adding sensors, motors, displays, and other modules to your Arduino Uno. Some common projects include:.As you gain experience, you can explore more complex projects by integrating sensors, actuators, and communication protocols. The vast Arduino community provides countless resources and examples to help you on your journey into the world of electronics and embedded programming.

Programming plays a pivotal role in the development and operation of IoT-based fertilizer pouring systems using electric vehicles. It encompasses various aspects, from controlling the vehicle's movement and fertilizer dispensing to data analysis and the remote monitoring. Programming algorithms for GPS-based navigation, path planning, and obstacle avoidance to enable the electric vehicle to move autonomously through the fields. Implementing code to control the vehicle's steering and speed based on GPS coordinates, sensor data, and pre-programmed routes. Programming the system to accurately dispense the required amount of fertilizer based on real-time sensor data (soil moisture, nutrient levels, etc.) and pre-defined schedules.

3.4 Input And Output

Digital Input Pins:

These pins can read either a HIGH (5V) or LOW (0V) signal.

They can be used to read the state of switches, buttons, sensors, or other digital devices.

Common examples include reading the state of a push button or a digital sensor like a temperature sensor. Note that this map is for the DIP-package chip. The Arduino Mini is grounded upon a lower physical IC package that includes two redundant ADC legs, which are not available in the DIP-package Arduino executions.

Each of the 14 digital legs on the Uno can be used as an input or affair, using pinMode(), digitalWrite(), and digitalRead() functions. They operate at 5 volts. Each leg can give or admit 20 mama as recommended operating condition and has an internal pull- up resistor(dissociated by dereliction) of 20- 50k ohm. A outside of 40Ma is the value that must n't be exceeded on any I/ O leg to avoid endless damage to the microcontroller.

- **AREF**. Reference voltage for the analog inputs. Used with analogReference().
- **Reset**. Bring this line LOW to reset the microcontroller. Generally used to add a reset button to securities which block the bone on the board.

Figure 2. Atmega168 Pin Mapping

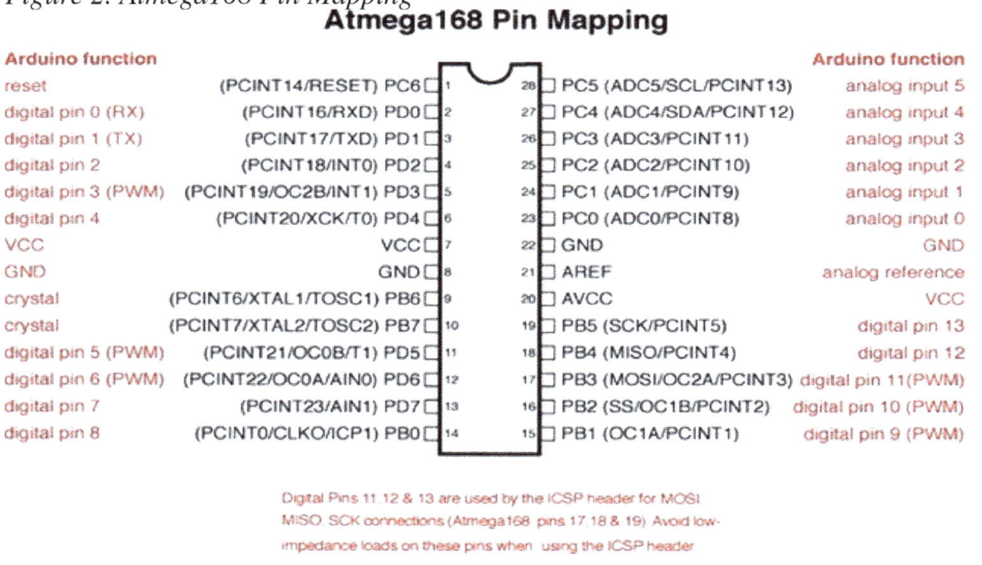

Digital Output Pins:

These pins can output either a HIGH (5V) or LOW (0V) signal.

They can be used to control the state of LEDs, motors, relays, and other digital devices.

For example, you can use a digital output pin to turn an LED on or off or control the speed of a motor.

Figure 3. H-Bridge Circuit

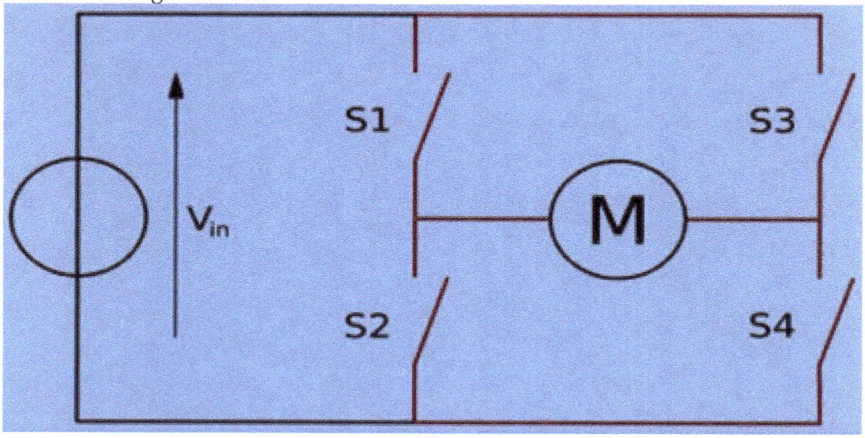

Switching Transistors: It typically consists of four transistors (often MOSFETs) arranged in a bridge configuration. Current Flow: By selectively switching these transistors on and off, the circuit can control the direction of current flow through the motor.

H-bridge is a type of electronic circuit that enables bidirectional control of a DC motor's direction of rotation. It's named for its resemblance to the letter "H" when diagramed.

Figure 4. Operation of the H-Bridge

S1	S2	S3	S4	Operation
1	0	0	1	Motor moves right
0	1	1	0	Motor moves left
0	0	0	0	Motor free runs
0	1	0	1	Motor brakes
1	0	1	0	Motor brakes
1	1	0	0	Short Power Supply
0	0	1	1	Short Power Supply
1	1	1	1	Short Power Supply

L293D Motor Driver IC

L293D IC is a typical Motor motorist IC which allows the DC motor to drive on any direction. This IC consists of 16- legs which are used to control a set of two DC motors presently in any direction. It means, by using a L293D IC we can control two DC motors. As well, this IC can drive small and quiet big motors.

This L293D IC works on the introductory principle of H- ground, this motor control circuit allows the voltage to be flowing in any direction. As we know that the voltage must be change the direction of being suitable to rotate the DC motor in both the directions. Hence, H- ground circuit using L293D Ics are perfect for driving a motor. Single L293D IC consists of two H- ground circuits inside which can rotate two DC motors independently. Generally, these circuits are used in robotics due to its size for controlling DC motors.

Figure 5. L293D IC Pin Configuration

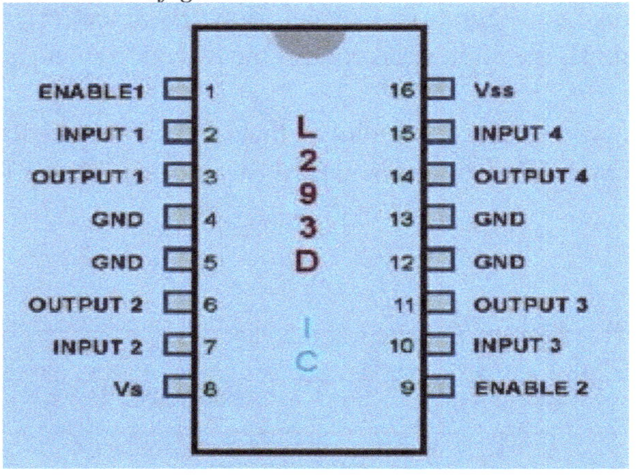

H Bridge Motor Control Circuit Using L293D IC

The IC LM293D consists of 4-i/p legs where, pin2 and 7 on the left side of the IC and leg 10 and 15 on the right side of the IC. Left input legs on the IC will control the gyration of a motor.

Figure 6. H-bridge Motor Circuit with L293D IC

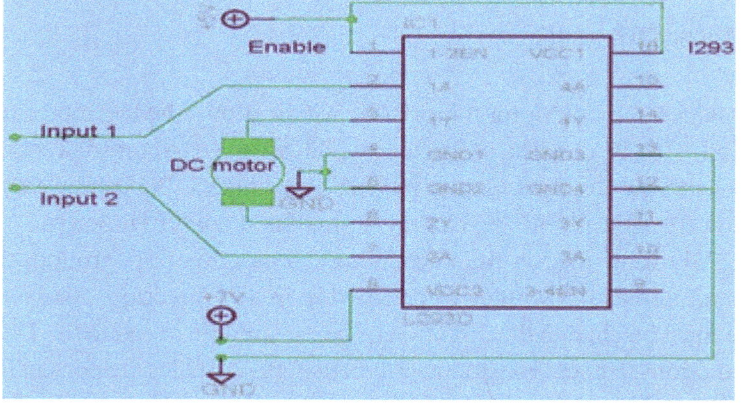

The L293D is a versatile integrated circuit that simplifies motor control by implementing the H-bridge configuration. This configuration allows for bidirectional control of DC motors, enabling them to spin in both forward and reverse directions.

By strategically switching the polarity of the voltage applied to the motor, the L293D effectively controls its rotation. Then, the motor is connected across side and right i/p for the motor on the right hand side.

This capability is crucial in various applications, including robotics, where precise motor control is essential for navigation and manipulation

- he L293D can control two DC motors simultaneously.
- Each motor requires a power supply of 4.5V to 36V.
- The maximum output current per channel is 600mA.
- To enable a motor, both the enable pin and the input pins must be set to appropriate

Direct Current (DC) Motor

A DC motor is an electromechanical device that converts direct current (DC) electrical energy into mechanical energy through the interaction of magnetic fields. It operates based on the principle that a current-carrying conductor placed in a magnetic field experiences a force, known as Lorentz force.

Basic Components of a DC Motor:

1. Armature: The rotating part of the motor that houses the windings.
2. Field Windings: Stationary coils that produce a magnetic field.
3. Commutator: A segmented ring that reverses the current direction in the armature windings.
4. Brushes: Carbon brushes that make electrical contact with the commutator...

Figure 7. Brushless DC motors work

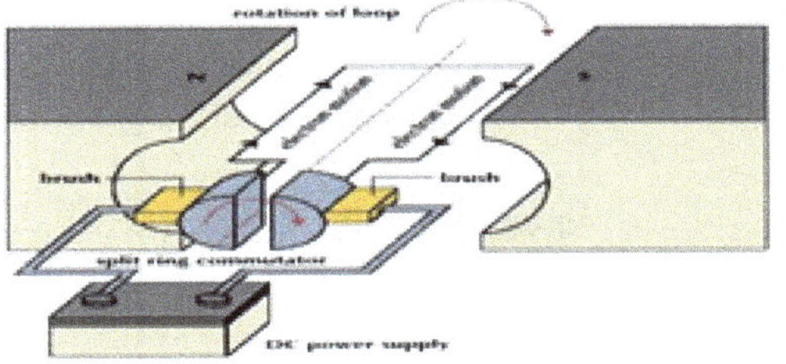

1. Current Flow: When direct current is supplied to the armature windings, it creates an
   electromagnetic field.

2. Interaction with Magnetic Field: This electromagnetic field interacts with the magnetic field
produced by the field windings.
3. Torque Generation: This interaction generates a torque, causing the armature to rotate.

Connecting DC Motor With Microcontroller:

While microcontrollers are powerful, they have limitations in terms of the amount of current they can directly supply. Directly connecting a DC motor to a microcontroller's pins can damage the microcontroller Solution: Using a Motor Driver ICTo overcome this limitation, we use a motor driver IC. This IC acts as an interface between the microcontroller and the motor, allowing the microcontroller to control the motor's speed and direction.

Dual H-Bridge Configuration: Enables bidirectional control of two DC motors Input Voltage Range: 4.5V to 36V, suitable for various power supply voltages. Output Current Capability:Each output can deliver up to 600mA of current. Input Logic Compatibility Compatible with both TTL and CMOS logic levels.

Figure 8. DC motor interfaced with L293D microcontroller

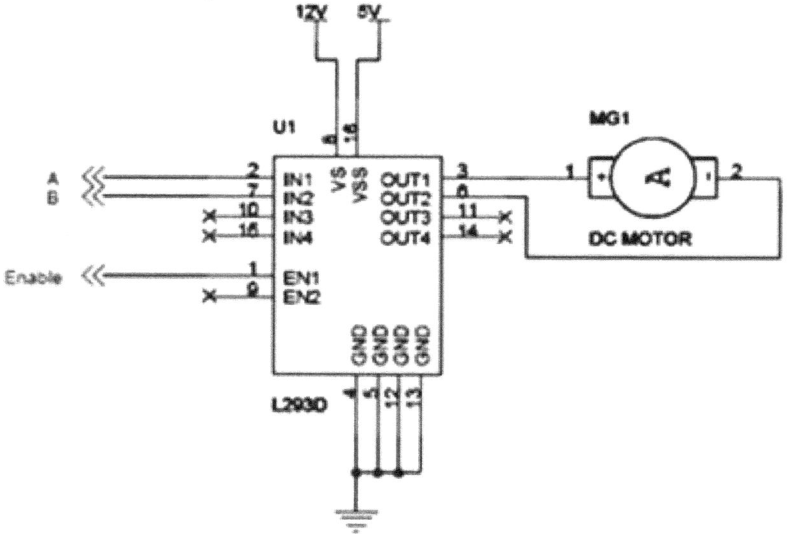

To connect a DC motor with a microcontroller, a motor driver or H-bridge circuit is used to handle the motor's higher voltage and current requirements. Directly connecting a motor to the microcontroller's GPIO pins is not feasible, as these pins cannot provide sufficient power.

A popular choice is the L298N motor driver module, which allows direction and speed control using digital and PWM signals from the microcontroller. The motor driver is powered by an external power supply, while the microcontroller controls the motor's behavior by setting the direction pins (e.g., IN1 and IN2) and adjusting speed using a PWM signal on the enable pin (ENA).

For example, setting IN1 high and IN2 low makes the motor rotate forward, while reversing these signals changes the direction. This setup ensures safe and efficient motor control for various applications like robotics, automation, and IoT projects.

Microcontroller: The brain of the operation. Popular choices include Arduino, Raspberry Pi, or 8051.

DC Motor: The actuator you want to control.

Power Supply: A voltage source suitable for both the microcontroller and the motor.

DC Motor Equations

V = E + IaRa

Where:
V: Applied voltage to the motor
E: Back EMF (electromotive force) generated by the motor
Ia: Armature current
Ra: Armature resistance
Torque Equation:

T = KφIa

Where:
T: Torque produced by the motor
Kφ: Motor constant (depends on the magnetic field and armature winding)
a: Armature current
Back EMF Equation:

E = K φ ω

Where:
E: Back EMF
K φ: Motor constant

ω: Angular speed of the

Speed Equation:

The back EMF is proportional to the motor speed:

*E = k e * ω*

Where *k e* is the EMF constant and *ω* is the angular speed.

Rearranging, motor speed can be expressed as:

*ω = (V – I a * R a) / k e*

This equation shows that speed decreases as the load increases (due to higher armature current and voltage drop).

In summary, these equations highlight the relationships between voltage, current, speed, and torque, allowing precise control and analysis of DC motor performance.

Types of DC Motors

Geared DC Motors

Geared motors tend to reduce the speed of the motor but with a corresponding increase in necklace. This property comes in handy, as DC motors can rotate at pets much too fast for an electronic device to makes use of. Geared motors generally correspond consist of a DC encounter motor and a gearbox attached to the shaft. Motors are distinguished as a geared by two connected units. It has numerous operations due to its cost of designing, reduces the complexity and constructing operations similar as artificial outfit, selectors, medical tools and robotics.c

Speed Reduction in Geared DC Motor

Figure 9. Speed reduction in geared DC motor

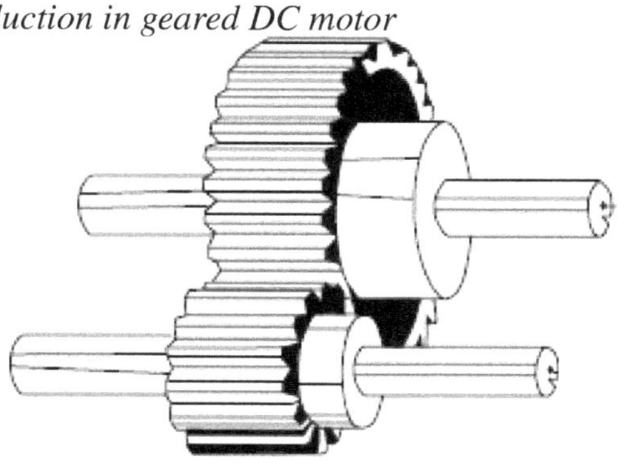

The gear reduction mechanism is a crucial component that allows for significant reductions in motor speed while increasing torque Here are the primary methods for speed reduction in geared DC motors Simple, robust gears that transmit power between parallel shafts.

Speed reduction in a geared DC motor is achieved by incorporating a gearbox (gear reduction mechanism) into the motor assembly. This mechanism reduces the rotational speed of the motor's output shaft while simultaneously increasing its torque. The process is based on the principles of mechanical advantage provided by gear ratios.

3.5 Series DC Motor

A series DC motor is a type of direct current motor where the field windings (stator) are connected in series with the armature winding (rotor). This configuration means that the same current flows through both the field and the armature. The performance characteristics of the series DC motor make it ideal for applications requiring high starting torque and variable speed control.

Figure 10. Series motor

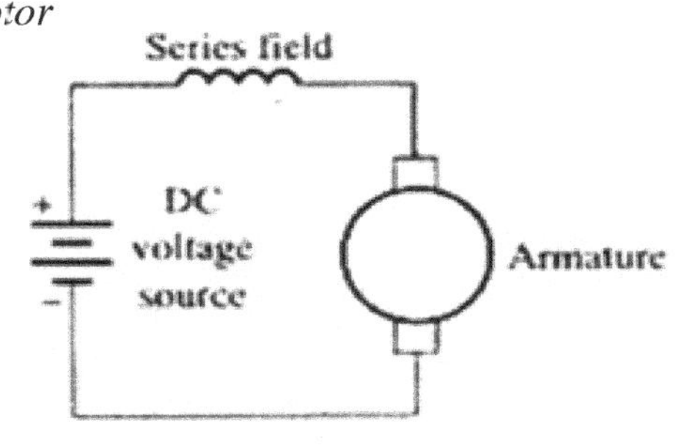

Due to the strong magnetic field created by the high initial current, series DC motors can produce very high starting torque. This makes them suitable for heavy load applications. The speed of a series DC motor decreases as the load increases. This is because the current through the motor increases with the load, which strengthens the magnetic field and reduces speed.

Advantages of Series Motor

- Simple Construction
- Designing is easy
- conservation is easy
- Cost effective

Ultrasonic Sensor Sound Waves

Robotics: Obstacle avoidance, line following, and object detection Automotive: Parking sensors, blind spot detection, and automatic braking systems.Industrial Automation: Level sensing, distance measurement, object detection in manufacturing processes. Consumer Electronics: Remote controls, gaming consoles, and smart home devices.

Thickness of material = Material sound velocity X Time of Fight

Transmission: The sensor emits a burst of ultrasonic sound waves.

Reception: The sensor receives the reflected sound waves (echoes).

Ultrasonic Sensor

An ultrasonic sensor is a device that uses sound waves to measure the distance to an object. It operates by emitting ultrasonic waves (high-frequency sound waves, typically above 20 kHz) and measuring the time it takes for the waves to bounce back after hitting an object. This time is used to calculate the distance between the sensor and the object using the speed of sound

.High- frequence sound swells reflect from boundaries to produce distinct echo patterns.

Figure 11. Ultrasonic sensor

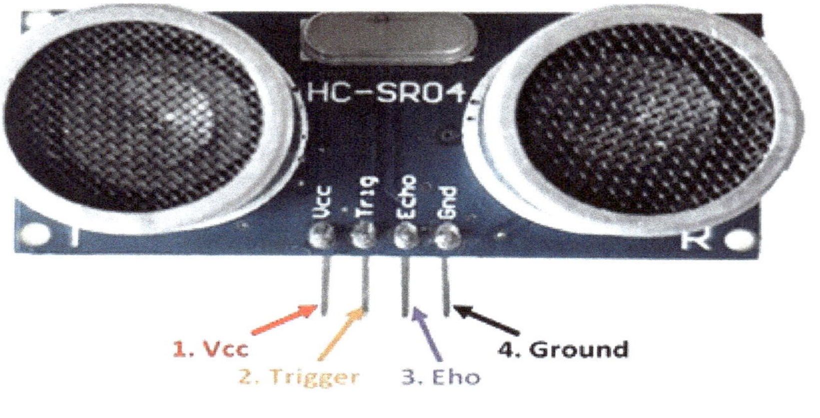

HC-SR04 Ultrasonic Sensor – Working

The HC-SR04 is a popular ultrasonic sensor module widely used in various projects due to its

simplicity and affordability.

How it Works:

1. Trigger Pulse: A short trigger pulse is sent to the sensor's trigger pin.
2. Ultrasonic Pulse Transmission: The sensor's transmitter emits a high-frequency ultrasonic pulse.
3. Echo Reception: The ultrasonic pulse travels through the air and reflects off an object. The sensor's receiver detects the reflected echo.
4. Time Measurement: The sensor measures the time taken for the pulse to travel to the object and back.,

Figure 12. Ultrasonic wave

Now, to calculate the distance using the below formulae, we should know the ower Supply: Connect the Vcc and GND pins of the sensor to the 5V and GND pins of the microcontroller, respectively. Trigger Signal: Use a digital output pin of the microcontroller to generate a short pulse (typically microseconds) to the Trig pin of the sensor.. Echo Signal: Connect the Echo pin of the sensor to a digital input pin of the microcontroller Timing Measurement: The microcontroller measures the pulse width of the echo signal to calculate the distance.

Applications

Ultrasonic sensors have a wide range of applications across various industries. Here are some of the most common applications:

Robotics and Automation

Obstacle Avoidance: Detecting obstacles in the robot's path to prevent collisions.

Line Following Detecting lines on the ground for autonomous navigation.

Proximity Sensing: Detecting objects within a certain distance for precise maneuvering.

3.6 DHT11 Sensor

The DHT11 detector is a compact, low- cost device designed to measure temperature and moisture situations in real- time. It operates on a digital communication protocol, making it easy to affiliate with microcontrollers and single- board computers like Arduino and Raspberry Pi.

Temperature Dimension: The DHT11 detector can measure temperatures within a range of 0 °C to 50 °C(32 °F to 122 °F) with an delicacy of ± 2 °C. It employs a thermistor to descry temperature changes and converts the analog signal into a digital format, making it suitable for colorful temperature monitoring operations.

Figure 13. DHT11 sensor

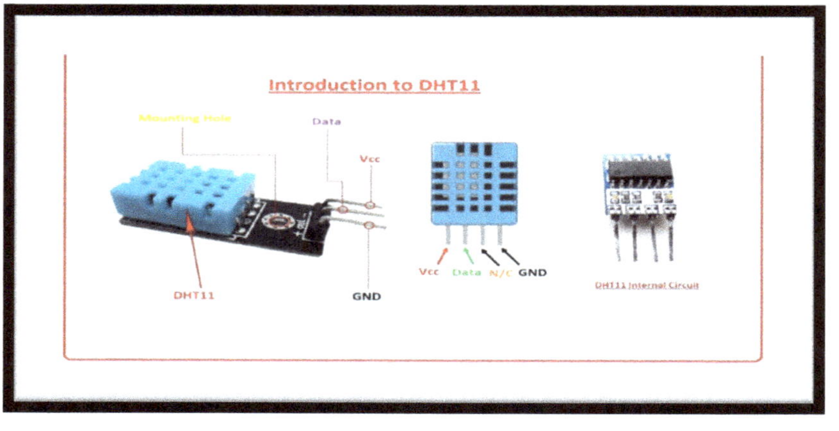

Moisture dimension: This detector can measure relative moisture(RH) in the range of 20 to 80 with an delicacy of ± 5. It uses a capacitive moisture detector to descry changes in moisture situations, offering precious data for moisture-sensitive operations similar as rainfall monitoring, HVAC control, and husbandry.

Power Conditions: This detector operates on a voltage force ranging from 3.3 V to 5V DC, making it compatible with utmost microcontroller development boards. It has a low power consumption, drawing only around 1 - 2mA during active dimension.

Response Time: The DHT11 detector has a fairly slow response time, generally taking about 2 seconds to give streamlined temperature and moisture readings. This detention should be considered when designing systems that bear real- time data.

Continuity and Trustability: While the DHT11 is cost-effective and simple to use, it may not be suitable for harsh or industrial environments due to its limited temperature range and accuracy. For more demanding applications, users may opt for more advanced sensors like the DHT22 or other high-precision sensors

How it Works:

1. Input Signal A control signal, typically a PWM (Pulse-Width Modulation) signal, is sent to the servo motor.
2. Signal Processing: The servo's internal circuitry decodes the PWM signal to determine the desired angular position.

3.Motor Rotation: The motor rotates to the specified position.

4. Potentiometer Feedback: A potentiometer, connected to the output shaft, provides feedback to the control circuitry..

Figure 14. Servo motor SG-90

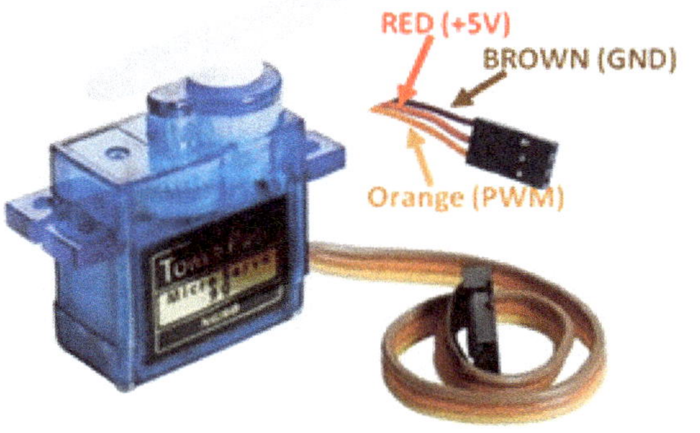

Specifications:

- Weight: 9 g

- Dimension: 22.2 x 11.8 x 31 mm approx.
- Stall torque: 1.8 kgf·cm
- Operating speed: 0.1 s/60 degree
- Operating voltage: 4.8 V (~5V)
- Dead band width: 10 µs
- Temperature range: 0 ºC – 55 ºC

Utmost servo motors have the following three connections 25

- Black/ Brown ground line.
- Red power line(around 5V).
- unheroic or White PWM line.

In this trial, we will connect the power and ground legs directly to the Arduino 5V and GND legs. The PWM input will be connected to one of the Arduino's digital affair legs

Table 2. Wire configuration:

Wire Number	Wire Colour	Description
1	Brown	Ground wire connected to the ground of the system
2	Red	Powers the motor typically +5v is used
3	Orange	PWM signal is given in through this wire to drive the motor

Applications of Servo Motors:

Robotics: Used in robotic arms, grippers, and other robotic mechaniss.Model Cars and Drones:*

Controlling steering, throttle, and other functions.

Industrial Automation: Used in assembly lines, packaging machines, and other automated systems.

Medical Devices: Used in prosthetic limbs, surgical robots, and medical equipment.

Toys and Hobbies: Used in remote-controlled toys and hobby projects.

Figure 15. Servo terminals

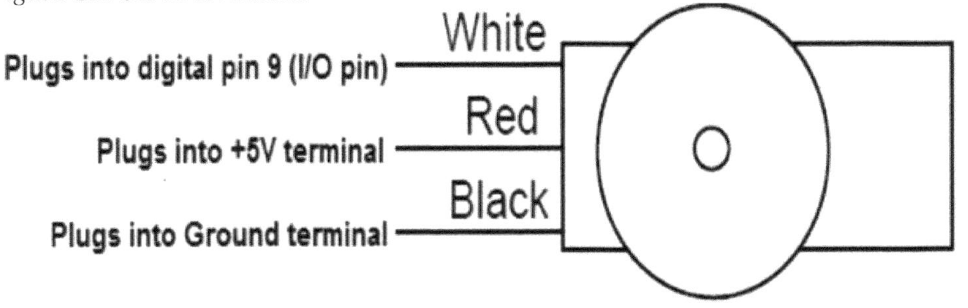

Advantages:

Precise Control: Can be accurately positioned to a specific angle.
High Torque: Can deliver significant torque, especially with gear reduction
Versatile: Can be used in a wide range of applications
Compact Size: Available in various sizes to suit different needs.

Disadvantages:

Limited Speed: Compared to other motor types, servo motors may have lower maximum speeds
Power Consumption: Can consume more power, especially when driving heavy loads.

Liquid Crystal Display

How LCDs Work:

1. Backlight:A backlight, typically an LED or fluorescent lamp, illuminates the LCD panel.
2. Polarizers: Polarizing filters align the light waves in a specific direction.
3. Liquid Crystal Layer: Liquid crystals, when subjected to an electric field, can twist the
   polarization of light passing through them.
4. Color Filter: A color filter layer determines the color of each pixel.
5. Front Polarizer A final polarizer filters the light, allowing only certain orientations to pass through.

Figure 16. LCD display

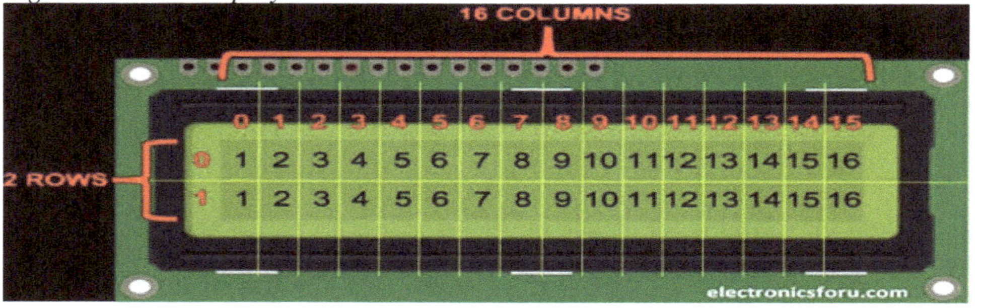

Types of LCDs:

TN (Twisted Nematic) LCD: The most common type, offering a fast response time but limited

viewing angles.

PS (In-Plane Switching) LCD Provides wider viewing angles and better color accuracy.

4. CALCULATION

$$\begin{cases} Q = \frac{sgt_1^2}{2} \cdot \frac{k}{2} \end{cases}$$

$$k = \frac{4/3\pi\left(\frac{d_0}{2}\right)^3}{2\left(\frac{d_0}{2}\right)^3}$$

$$t_1 = \frac{\theta}{\omega}$$

$$S = \frac{\theta}{360^0}(R^2 \Pi - r^{2\pi}) \tag{1}$$

$$Q_2 = S\left(v_1 t_2 + \frac{gt_2^2}{2}\right)k$$

$$t_2 = \frac{60 - \theta}{\omega}$$

$$v_1 = \frac{H - H_1}{t_1}$$

$$R_2^2 H_1 = R_1^2 H - \frac{3Q_1}{\pi}$$

$$\frac{R_1}{H} = \frac{R_2}{H_1}$$

$$Q_3 = s\left[v_2 t_3 + \frac{g t_3^2}{q}\right] \cdot \frac{k}{2}$$

$$t_1 = t_3 = \frac{\theta}{\omega}$$

$$v_2 = \frac{H_1 - H_2}{t_2}$$

$$R_3^2 H_2 = R_2^2 H_1 - \frac{3Q_2}{\pi}$$

$$\frac{R_2}{H_1} = \frac{R_3}{H_2}$$

$$Q_L = 3\rho\left(Q_1 + Q_2 + Q_3\right) \tag{2}$$

$$\theta - {}^{\pi}\!/_{6} = \beta - \alpha = arccos\left|\frac{\mathrm{d}_1^2 + L_2^2 + \mathrm{d}_2^2 - L_2^2}{2d_1\sqrt{L_1^2 + d_2^2}}\right| - arccos\left(\frac{\mathrm{d}_2}{\sqrt{L_1^2 + d_2^2}}\right) L \geq 0 \tag{3}$$

$$\theta - {}^{\pi}\!/_{6} = \beta + \alpha = arccos\left|\frac{\mathrm{d}_1^2 + L_2^2 + \mathrm{d}_2^2 - L_2^2}{2d_1\sqrt{L_1^2 + d_2^2}}\right| + arccos\left(\frac{\mathrm{d}_2}{\sqrt{L_1^2 + d_2^2}}\right) L_1 < 0 \tag{4}$$

$$\theta = 0.833L + 6.667 \tag{5}$$

$$L = (20\mu + 4)\Delta t \tag{6}$$

$$Q = \left(D_0 - k' d_0\right)^{3/2} \tag{7}$$

$$Q = \frac{\pi}{6}\rho\ \sqrt{g}\left(D_0 - k' d_0\right)\frac{5}{2}\left(\frac{1 - \cos^{3/2}\varnothing}{\sin^{5/2}\varnothing}\right) \tag{8}$$

$$Q = A(L, -k'd')\left(W_{outlet-k'd'}\right)^{3/2} \tag{9}$$

$$Q = C_\rho \sqrt{g}\left(D_0 - k' d_0\right)^{5/2} \tag{10}$$

$$W_\gamma = M\rho\sqrt{g}\left(\frac{D' - k'' \mathrm{d}_0}{n'}\right)^{2.5}\left[t_1 + t_2\right] \tag{11}$$

$$D' = \sqrt{R^2 - r^2} \tag{12}$$

$$n' = \frac{360^0}{\theta} \tag{13}$$

$$t_1 = Tt'\left(\frac{\theta}{180}\right)\left[\frac{1}{24 - \left(\frac{\theta}{5}\right)\varepsilon}\right] \tag{14}$$

$$t_2 = Tt'\left(\frac{60^0 - \theta}{360^0}\right) \tag{15}$$

$$Q_h = W_r \times n \tag{16}$$

$$v = 0.00926 - 0.0000029 \;;$$

$$Q_s = \frac{Q_h}{60 \times 60} \;;$$

$$Q_1 = \frac{1}{v} Q_s; z = \frac{l}{b}; Q_z = Q_1 z \tag{17}$$

Figure 17. Project output

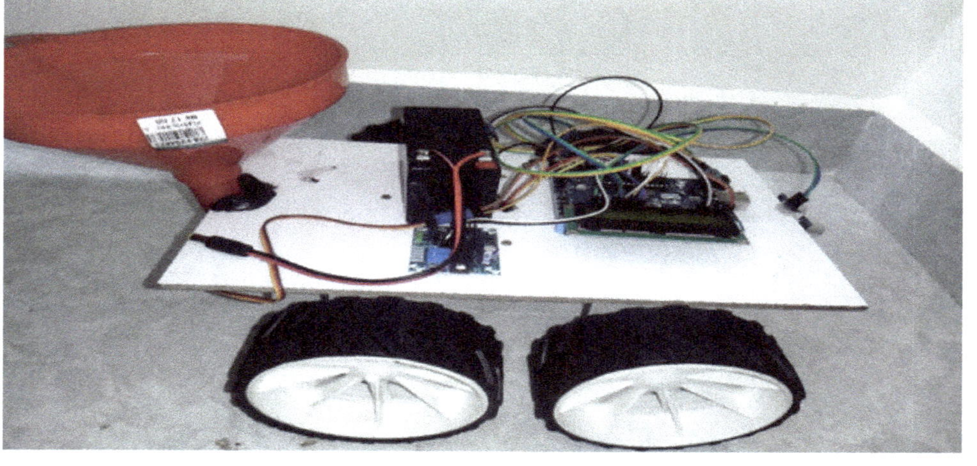

5. FUTURE ENHANCEMENT

5.1 Applications

Precision Agriculture The robot can be used in perfection husbandry to optimize toxin operation grounded on real- time environmental data. This targeted approach improves crop health and yield while minimizing resource waste and environmental impact.

Large- Scale Farm Management It's ideal for large agrarian operations where homemade spraying is hamstrung and expensive. The robot's independent capabilities and remote control via IoT streamline field operation, reduce labour conditions, and insure harmonious operation across expansive areas.

Remote Monitoring: Real-time monitoring and control of fertilizer application via smartphones or computers.

Cost Efficiency: Reduction in fertilizer wastage, leading to cost savings and sustainable practices.

Environmental Protection: Minimized over-fertilization reduces runoff and protects ecosystems.

Vehicle Deployment: Combining Vehicle with IoT systems for aerial fertilizer application.

5.2 Future Enhancement

Unborn advancements for this independent toxin sprayer could concentrate on expanding its capabilities to ameliorate perfection, effectiveness, and rigidity in different agrarian surroundings. Integrating fresh detectors similar as soil humidity, pH, and light detectors would allow the robot to gather further comprehensive data, enabling more accurate and customized toxin operations grounded on specific crop requirements and soil conditions. The addition of a GPS system for real- time geolocation could further ameliorate navigation delicacy, allowing the robot to followpre-defined paths more precisely and perform tasks like automated mapping of the field for effective toxin content. To increase the robot's autonomy and effectiveness, advanced machine learning algorithms could be integrated for better decision- timber, similar as determining optimal scattering patterns or feting crop health issues, which could prompt corrective conduct or pest discovery. also, enhancing the robot's battery life and incorporating solar panels for recharging would extend its functional time in the field, reducing time-out. bettered wireless communication and integration with pall- grounded ranch operation platforms could allow growers to cover multiple robots in real time, manage operations ever, and dissect performance data for farther optimization. Future enhancements of fertilizer pouring in agriculture could include integrating advanced AI and machine learning algorithms to analyze real-time data from IoT sensors more effectively, enabling even more precise and predictive fertilizer application. Innovations like drones or autonomous robots equipped with smart sensors could further automate and optimize fertilizer distribution across large fields. Additionally, incorporating renewable energy sources, such as solar-powered IoT devices, would enhance sustainability. Enhanced interoperability with other smart farming technologies, such as weather forecasting systems and crop monitoring platforms, could provide holistic farm management, reducing waste, improving yield quality, and addressing environmental concerns more comprehensively.

5.3 Conclusion

In conclusion, the development of this independent toxin sprayer represents a significant advancement in the field of perfection husbandry, offering a cost-effective, effective, and environmentally conscious result for toxin operation. By integrating environmental monitoring, independent navigation, and handicap discovery, the system ensures that diseases are applied precisely when and where they're demanded, optimizing resource operation and minimizing waste. The use of the Arduino Uno as the main regulator and ESP8266 for pall communication enhances the robot's functionality, enabling real- time monitoring and control through the IoT. This position of robotization not only reduces labor costs but also promotes sustainable husbandry practices by precluding overuse of diseases and minimizing environmental impurity. With its capability to acclimatize to varying field conditions and operate autonomously or under stoner control, the robot is protean enough for use in different agrarian settings. The overall design addresses crucial challenges faced by growers, including high outfit costs and the complexity of being agrarian technology. By furnishing a practical, affordable, and scalable result. The simplest way to control a DC motor is to turn it on and off. The microcontroller can send a digital signal to the motor driver to switch the motor on or off. In conclusion, the development of this independent toxin sprayer represents a significant advancement in the field of perfection husbandry, offering a cost-effective, effective, and environmentally conscious result for toxin operation. By integrating environmental monitoring, independent navigation, and handicap discovery, the system ensures that diseases are applied precisely when and where they're demanded, optimizing resource operation and minimizing waste. The use of the Arduino Uno as the main regulator and ESP8266 for pall communication enhances the robot's functionality, enabling real-time monitoring and control through the IoT. This position of robotization not only reduces labor costs but also promotes sustainable husbandry practices by precluding overuse of diseases and minimizing environmental impurity. With its capability to acclimatize to varying field conditions and operate autonomously or under stoner control, the robot is protean enough for use in different agrarian settings. The overall design addresses crucial challenges faced by growers, including high outfit costs and the complexity of being agrarian technology. By furnishing a practical, affordable, and scalable result.will spin. Many DC motors can be made to rotate in both directions. This is achieved by reversing the polarity of the voltage applied to the motor. The microcontroller can control the direction by sending appropriate signals to the motor driver. An IoT-based fertilizer pouring system optimizes agricultural practices by leveraging sensors and automated mechanisms to deliver precise amounts of fertilizer based on real-time soil and crop conditions. This technology enhances resource efficiency, reduces environmental impact, and improves crop yield while minimizing labor-intensive tasks. By integrating IoT devices with data analytics, farmers can achieve sustainable farming practices and adapt to dynamic agricultural needs, fostering better productivity and eco-friendly outcomes.

REFERENCES

Brooks, R. A. (1986). A robust layered control system for a mobile robot. *IEEE Journal on Robotics and Automation*, *RA-2*(1), 14–23. DOI: 10.1109/JRA.1986.1087032

Gat, E. (1998). On three-layer architectures. In *Artificial intelligence and mobile robots* (Vol. 195, Article 210).

Kaelbling, L. P., Littman, M. L., & Moore, A. W. (1987). An architecture for intelligent reactive systems. In *Reasoning about actions and plans* (pp. 395–410). Elsevier. DOI: 10.1016/B978-0-934613-30-9.50019-6

Kootstra, G., Bender, A., Perez, T., & van Henten, E. J. (2020). *Robotics in agriculture*. Springer., DOI: 10.1007/978-3-642-41610-1_43-1

Martin, T., Hossard, C., Therond, O., Baret, F., Colbach, N., Messéan, A., & Munier-Jolain, N. (2022). Robots and transformations of work in farm: A systematic review of the literature and a research agenda. *Agronomy for Sustainable Development*, *42*(4), 66. Advance online publication. DOI: 10.1007/s13593-022-00796-2

Mehra, A. (2020). *Agricultural robots market worth USD 20.3 billion by 2025*. MarketsandMarkets. https://www.marketsandmarkets.com/MarketReports/agricultural-robot-market-173601759.html

Murphy, R. R. (2019). *Introduction to AI robotics*. MIT Press.

Niroui, F., Zhang, K., Kashino, Z., & Nejat, G. (2019). Deep reinforcement learning robot for search and rescue applications: Exploration in unknown cluttered environments. *IEEE Robotics and Automation Letters*, *4*(2), 610–617. DOI: 10.1109/LRA.2019.2891991

APPENDIX

Table 3. List of Abbreviation

Acronyms	Description
PA	Precision Agriculture
TSP	Traveling Salesman Problem
RIZE	Robot Interfaces From Zero Experience
SAG	Stochastic-Vertex-Cost Aisle Graph
NBA-P	Next-Best-Action Planning
DFU	Device Firmware Update
ADC	Analog-to-digital converter
AREF	Analog Reference
RH	Relative humidity
GPS	Global Positioning System

Chapter 8
Web-Based Diagnostic Tool for Pulse Crop Disease Identification Using Convolutional Neural Networks

Ketan Bodhe
http://orcid.org/0000-0002-6357-6627
G.H. Raisoni College of Engineering, Nagpur, India

Kamlesh Ashokrao Kalbande
http://orcid.org/0000-0003-2705-1732
G.H. Raisoni College of Engineering, Nagpur, India

Dhananjay Bhagat
http://orcid.org/0009-0009-1100-3219
Department of Computer Science and Engineering (AI), Vishwakarma Institute of Technology, Pune, India

Bhagyashree Hambarde
http://orcid.org/0000-0002-7331-5164
St. Vincent Pallotti College of Engineering and Technology, Nagpur, India

Pranay Saraf
http://orcid.org/0000-0003-4169-7118
G.H. Raisoni College of Engineering, Nagpur, India

Asakti Rautkar
G.H. Raisoni College of Engineering, Nagpur, India

Ashvini Ganeshsingh Bais
http://orcid.org/0009-0000-9302-3604
G.H. Raisoni College of Engineering, Nagpur, India

DOI: 10.4018/979-8-3373-2429-6.ch008

ABSTRACT

Plant diseases threaten food security, especially in pulse crops essential for sustainable agriculture. This study explores Convolutional Neural Networks (CNNs) for rapid detection of major diseases such as Ascochyta blight, Fusarium wilt, and phytoplasma infections, which cause severe losses in countries like Ethiopia, India, and Taiwan. Using advanced imaging, we propose a web-based diagnostic tool to empower farmers with timely, accurate disease identification, enabling effective management decisions. This approach offers not just technological advancement but also economic sustainability for smallholder farmers by improving crop health and yields through early intervention and precision agriculture practices.

INTRODUCTION

Pulses, such as lentils, chickpeas, and beans, play a major role in sustainable agriculture as they construct nitrogen-fixing bacteria that will lead to soil fertility improvement. Legumes, as protein suppliers and main nutrient donors, are greatly needed mostly in countries abroad where people are accustomed to only a few varieties of foodstuffs (Ingole et al., 2023). They are the weighty and economically effective methods to nourish the required vitamins, especially when the animal products are not available. The planting of pulses, chickpeas, cowpeas, and mung beans is crucial in food security and sustainable agriculture. However, some crops are prone to leaf infestations that will cause a decline in the quality and quantity of the produce (Thakre et al., 2023).

Diseases like Ascochyta blight, Fusarium wilt, and phytoplasma infections are the most difficult challenges for farmers in Ethiopia, India, and Taiwan. In Ethiopia, where pulses are the main food source, these diseases can cause a huge loss of crops that are a family's source of income and nutrition. In India, many farmers grow crops that are high in protein and are of high quality but their plants are affected by diseases like Fusarium wilt and Ascochyta blight that might kill their whole yield. A majority of these farmers are those who work on small plots and have very minimal resources to fight the challenges that come with these diseases thus their livelihoods are very much at risk (Giradkar et al., 2023). The resolution of these issues requires the empowerment of the farmers and is crucial for the success and support of their societies. However, in Taiwan, the question of the effect of phytoplasma-induced diseases on pulse production still exists, making it difficult to evaluate the particular impact on the area's farming diversity. The good news is that the solution to these diseases in the firmly agricultural sector will curb the condition of food security for the region, lower consumer prices, and improve the welfare of the farmers worldwide (Kalbande et al., 2021).

A better understanding of the cause of these disorders, accurate diagnostic testing, and the right way of processing them are the main things to do to avoid their harmful effects (Tandekar & Dongre, 2023).

Old-fashioned methods of finding out diseases, for example, by human investigation and laboratory tests, are not enough when it comes to large-scale agriculture, especially in remote areas (Shende, Thakare, Agrawal et al, 2023). This problem is worsened by the existence of insect-borne viruses, which considerably affect the number of crops and their quality. Web-aided plant disease detection is the new thing in this industry, thanks to the impact of the latest advances in deep learning and computer vision; it gives various benefits including automation and highly improved diagnostics (Pillewan et al., 2023).

As a part of our conducted research, we have a plan to make a web-based tool for the diagnosis of plant diseases in pulse crops, like peas, lentils, black grams, etc., using CNNs. The main goal of this program is to educate farmers about the methods to take care of their crops to maintain health and high yields. Thus, they will be encouraged to develop new and improved crop management practices (Shende, Wyawahare, Thakare et al, 2023).

Crops that are mostly grown and their most common diseases will be the center of our interest. Among them, I will just mention some like Fusarium wilt Ascochyta blight, and phytoplasma infections (Gajbhiye & Dongre, 2013). Our main focus is the provision of the possibility of early disease detection and the efficient running of disease management to crop farmers and in this wise the development of a platform that is fully user-oriented and at the same time accessible to them. This way of trading with the farmers is expected to serve the ultimate purpose of providing the conditions of their livelihoods and also contributing process to food security in places that heavily rely on the production of such crops (Pandhare et al., 2021).

Table 1. Causes and Symptoms of Plant Disease

Name of Disease	Causes	Symptoms
Ascochyta Blight and Fusarium Wilt (Chickpeas)	Fungal pathogens.	Ascochyta blight causes dark necrotic lesions on leaves and pods, while Fusarium wilt leads to vascular discoloration and plant wilting.
Anthracnose, Leaf Crinkle, Powdery Mildew, and Yellow Mosaic (Black Gram)	Fungal pathogens and viruses.	Anthracnose results in leaf spotting, Leaf Crinkle causes leaf deformation, Powdery Mildew produces white fungal growth on leaf surfaces, and Yellow Mosaic causes mottled, discolored leaves
Macrophomina Leaf Blight (Mung Beans)	Fungal infection by Macrophomina phaseolina	Irregular, water-soaked lesions that merge into large necrotic areas, reduce photosynthesis and yield.
Phytoplasma-Induced Diseases (Mung Beans)	Phytoplasmas (intracellular pathogens)	Abnormal leaf growth, premature seed germination, and stunted development.
Rust Disease (Lentils)	Fungal pathogen *Uromyces fabae*	Small, orange-brown pustules form on the underside of leaves, leading to premature leaf drop, reduced photosynthesis, and stunted plant growth.

LITERATURE REVIEW

Review of Existing Tools

In recent years, advancements in techniques for the detection of plant disease have changed substantially and improved efficiency and accuracy. The ability to detect diseases is critical for managing crop yields and the sustainability of agriculture (Belay et al., 2022). These traditional methods typically begin with visual inspection, which is the basis of the assessment, but rarely are specific enough to provide the accuracy desired. The visual inspections depend on the expertise of a trained agronomist to physically and manually identify symptoms on the crops that can lead to mistakes and misunderstandings about specific plant diseases (Talasila et al., 2022).

These drawbacks have all been addressed, as visual assessments are now being paired with novel methods that combine technology to better predict visual assessments and provide more accuracy. Novel advancements in image processing and computational analysis have changed the norm of being able to deliver the rapid and exact identification of symptoms of disease. These advancements incorporate the use of algorithms, and machine learning to run images, thus also reducing human error (Yasin et al., 2023).

Field surveys provide additional invaluable information that reinforces the understanding of the frequency and severity of diseases. Moreover, techniques that involve molecular understanding of pathogens, such as Polymerase Chain Reaction (PCR) or Enzyme-Linked Immunosorbent Assay (ELISA) are both precise assays that provide against the organism to provide specific details when considering protocols for management (Khaire et al., 2023).

When combined these methods, work together to build a more integrative and methodical approach to plant disease detection which improves the accuracy of diagnoses and potential management skills for addressing plant diseases. A methodical approach allows farmers to respond timely to new and emerging threats to agriculture, thus protecting crops and contributing to food security (Antichi et al., 2023).

Figure 1. Exploring Disease Management in Pulse Crops

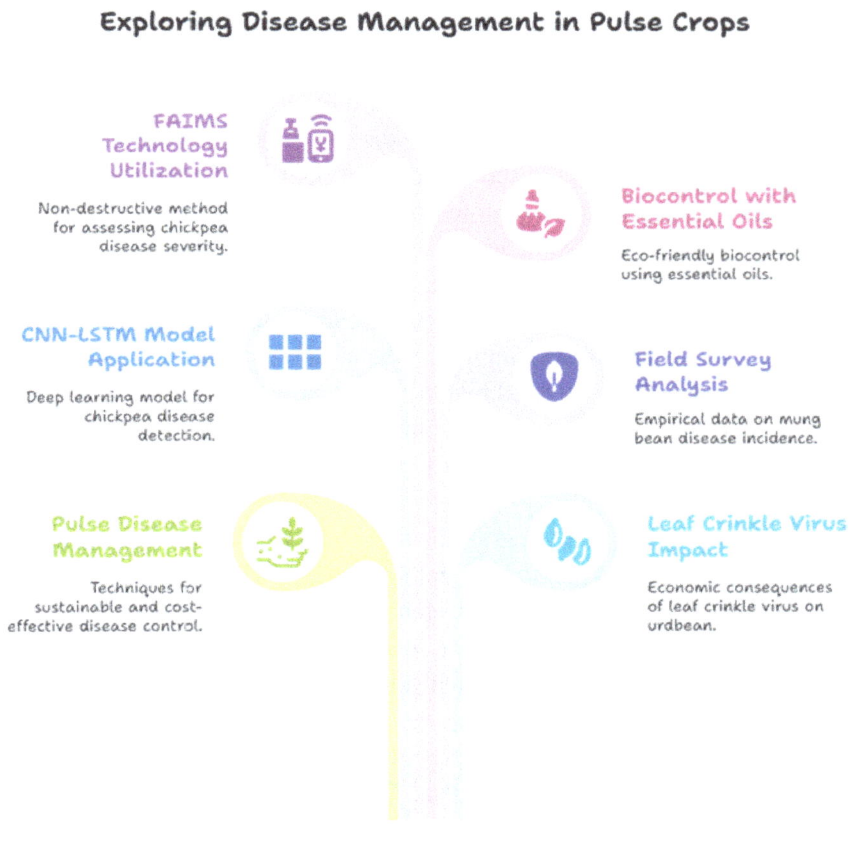

Study of Diseases

Research on disease in pulse crops is necessary for increased agricultural productivity and food security. This study considers several key areas of plant disease, its causal agents and symptoms, the progression of plant disease, and the impacts of the disease on yield and quality.

Disease in pulse crops is caused by a variety of pathogens, such as fungi, bacteria, and viruses. Accurately understanding the processes of infection and the conditions conducive to pathogen proliferation is crucial to developing strategies to manage diseases. Environmental factors, such as humidity, temperature, and soil type all play a role in disease outbreaks.

Properly identifying symptoms is important for diagnosing a problem and taking the necessary action. Symptoms can be leaves losing coloring and beginning to spot, wilting, and wilting or wilting of the entire plant. Diagnosing plants today is primarily done through visual inspections and knowledge from specialists, although newer diagnostic methods, with the advent of molecular assays and imaging, has improved the speed and accuracy of on-site disease diagnosis (Zhang et al., 2023).

Disease in turn takes effect based on different pathogens and the environments. Some diseases will display slower to infest plants and some diseases can grow rapidly through water, wind, and beyond an injured seed. Our understanding of the cause and the effects of infested and infected plants is vitally important in how and when appropriate practices and management actions can occur.

The impact of plant diseases can have substantial impacts on yield and quality, resulting in economic loss. Infected plants can have no or limited growth and could be affected if there is a low pod with limited seeds, which will affect harvests. Definition impacts can be converted with agronomic field studies and loads (Chen et al., 2021).

Management requires the integration of the use of disease-resistant varieties, modified cultural practices, and chemical applications. Detection is a process involving rounds of overwatch once interested in emergence. Active surveillance and monitoring are needed when taking action- management to mitigate outbreaks. Some highlighted research could be initiated with an assessment of systematic agricultural practices all with challenged by disease.

Machine Learning in Agriculture: Overview

The incorporation of machine learning technologies, especially Convolutional Neural Networks (CNNs), has revolutionized the way plant diseases are detected and regulated. In this overview, we will discuss some of the relevant research on CNN applications in the agricultural sector.

Successful Applications of Machine Learning Technologies

In particular, image classification technology has shown promising applications in classifying images of healthy and diseased or infected plants, while simultaneously showing accuracy with classifications of images according to the various diseases. The following details contain several of the research and successful implementations of the technology in the agriculture sector(Iqbal et al., 2019).

1. Image Classification and Taxonomy: Recent research has clearly shown how CNNs are better than traditional visual inspections of detecting plant diseases from images, producing high accuracy rates. CNNs, in particular, are excellent at processing pixels or visual data and have enabled researchers and plant pathologists to perform assessments in plants where visible symptoms may be present, and detect symptoms associated with particular pathogens.
2. Real-time Monitoring of Crops: Real-time monitoring applications are an exciting recent development. CNN applications have been reducing the time of identifying the presence of any disease outbreak. New methods of quickly identifying plant disease through CNNs could be successful in limiting crop damage.
3. Field Surveys: Farming practices can entrain data collection through automated technology by measuring symptom expression in unseen ways (Rana et al., 2016).
4. IoT: Some excitement is generated through smart farming solutions based on CNNs and IoT (Internet of Things) applications. Solutions that can monitor environmental conditions and potential disease outbreaks are of particular interest.

Challenges of Applying CNNs:

Although computer vision has been successful in various applications of agriculture technology, there are several challenges in the widespread application of CNNs for early disease detection in plants (Gautam et al., 2016).

Data Quality and Quantity: In general, CNN models require enough good-quality labeling data. In most cases, obtaining datasets, that cover different plant species and diseases, is challenging due to the scarcity of data.

Generalization: The generalization of CNN models that are trained on diverse datasets is hard because it may not work perfectly under different environmental conditions, different lighting conditions, or different plant varieties. CNN models need to be made robust and they should be able to adapt to various circumstances.

Interpretability: Humans, such as farmers or agronomists, should understand the prediction of plant diseases to trust automated systems. The amount of trust the user has in CNN based on trained systems is critical for uptake.

Computational Requirements: One of the biggest issues with machine learning in general is the computational resources. For many practical systems, including an Internet of Things (IoT) system of wireless node-embedded sensors, with the power of the smartphone to the center, providing real-time monitoring of plant disease, would require massive computational resources. Many farmers in the developing world have limited access to such technology.

Integration into Existing Practice: These machine learning systems are new to farmers, and their uptake is the essential component of the successful implementation of the imaging system for plant disease diagnosis. For the scalability of such an application, a computer-literate workforce and youth must replace the older workers who have just restarted following the introduction of the green revolution in the 1970s (Martini et al., 2018).

CNNs find a great application in plant disease detection. However, creating a reliable system for implementing machine learning is complicated given trends, the challenge of data quality and quantity, generalization, interpretability of results to the user, and the feasibility of forward implementation. There is a lot of potential in machine learning to enhance agricultural technology.

COMMON DISEASES IN PULSES

Pulse crops, consisting of lentils, pigeon peas, chickpeas, black gram, green gram, and moth beans, are very important for farming in India because they are nutritious, help improve soil health by adding nitrogen, and are used in crop rotation to keep the soil fertile and reduce pests. However, these crops are prone to various diseases that can significantly affect their yield and quality. In this particular section, we will be discussing some of the common diseases found in pulse crops, along with their stages, symptoms, physiological impacts, and visual indicators (Rashed et al., 2018).

A. Fusarium Wilt

Fusarium Wilt is a devastating soil-borne fungal disease caused by Fusarium oxysporum f. sp. ciceri. This disease attacks the plant's internal system which transports water and nutrients to the plant. Because of this, the needs of the plant aren't fulfilled resulting in the leaves to wither and turn yellow. It usually happens in places that are usually humid.

Initial Stage

Symptoms: Leaves at the lower end of the plant start yellowing and withering, often during the day, and during the night it recovers.

Plant Physiology: The fungus attacks the root system, blocking the xylem vessels, which makes it difficult for water and nutrients to reach the rest of the plant.

Visual Indicators: The yellowing of the leaves can be seen on the lining of the leaves and moves inward (Gossen et al., 2016).

Intermediate Stage

Symptoms: Yellowing moves to the upper region of leaves, while withering becomes permanent. The plant growth is stagnant point, and leaves may start to fall.

Plant Physiology: fungal colonies increase in the stem and root, causing an upsurge in vascular damage.

Visual Indicators: The color of leaves changes from green to completely yellow, and the withering is more prominent. If cut open the stem might show discoloration.

Advanced Stage

Symptoms: There is an occurrence of withering, yellowing, and browning of leaves. The plant eventually collapses and dies as the water and nutrient supply is obstructed.

Plant Physiology: there is a complete spread of fungus alongside vascular blockage in the plant.

Visual Indicators: Dead and dried leaves, stems, and overall plant collapse (Pattanayak et al., 2019).

B. Ascochyta Blight

Ascochyta Blight is caused by a group of fungal pathogens including Ascochyta rabiei (chickpeas), Ascochyta lentis (lentils), and Ascochyta pisi (peas). This disease can be identified by observing the spots on leaves and stems leading to reduced ability to perform photosynthesis and significant yield losses. The disease is seen in cool, wet conditions.

Initial Stage

Symptoms: There is the appearance of small, water-soaked lesions on leaves, and stems.

Plant Physiology: The entrance of germinated spores facilitates the infection into the plant tissue.

Visual Indicators: Small lesions might appear with small light rings.

Intermediate Stage

Symptoms: Black and Brown enlarged lesions appear with a concentric ring pattern. Deep lesions are seen on the stem and pods of the plant.

Plant Physiology: Cellular integrity is disrupted by fungal growth and toxin production.

Visual Indicators: Lesions with concentric rings, indicate advanced fungal colonization.

Advanced Stage

Symptoms: Lesions are merged because of severe infection which leads to extensive decay. Plants might lose their leaves, and the pods might dry up, leading to bad seed quality.

Plant Physiology: Serious tissue damage and minor infections.

Visual Indicators: Large dead areas, completely fallen leaves, and dried, shriveled pods.

C. Rust

Rust diseases in pulses are caused by several fungal pathogens, including Uromyces spp. and Phakopsora spp. These diseases are identified by the formation of rust-colored spots on leaves, stems, and pods, which weakens the plant and makes the photosynthesis process difficult. Rust usually appears in humid conditions and can spread rapidly in these conditions (Kalbande et al., 2021).

Initial Stage

Symptoms: Occurrence of small, white, or yellowish spots appear on the upper surface of the leaves.

Plant Physiology: Spore germination on leaf surfaces and initial infection.

Visual Indicators: Small, yellowish spots on leaves are seen.

Intermediate Stage

Symptoms: Spots become small bumps that release rust-colored powder and the bottom of the leaves also has these bumps.

Plant Physiology: Fungal growth inside the leaf tissue, spore production, and dispersal.

Visual Indicators: Raised orange to brown bumps are seen on both leaf surfaces.

Advanced Stage

Symptoms: Bumps merge damaging the leaves and severe infections result in early falling of leaves and weakening of plants.

Plant Physiology: The ability to perform photosynthesis is hindered due to extensive leaf tissue damage.

Visual Indicators: Large bumps that are merged, heavy falling of leaves, and severely weakened plants.

D. Powdery Mildew

Powdery Mildew is a common fungal disease in pulses caused by Erysiphe pisi and related species. It is seen as white, powdery fungal growth on the surfaces of leaves, stems, and pods, reducing the possibility of photosynthesis and compromising the health of the plant. Powdery Mildew grows well in dry, warm conditions with high humidity (Shende, Thakare, Agrawal et al, 2023)

Initial Stage

Symptoms: Occurrence of small, white, powdery spots on the upper surface of leaves.

Plant Physiology: Fungal spores are seen on the surface of the leaves with mycelial growth.

Visual Indicators: Small, white powdery spots can be seen.

Intermediate Stage

Symptoms: On a large surface area on the leaf spots can be seen that grow and merge. Leaves that are infected turn yellow and distorted.

Plant Physiology: Increased fungal biomass on leaf surfaces, hindering photosynthesis.

Visual Indicators: Large white, powdery patches, leaf yellowing and curling (Shende, Wyawahare, Thakare et al, 2023).

Advanced Stage

Symptoms: Severe infections create a thick, powdery coating on leaves, stems, and pods. Infected tissues may turn brown and die, this leads to leaves falling off and reduced pod formation.

Plant Physiology: Fungal growth completely covers the whole plant which causes severe stress on the plant.

Visual Indicators: A thick white layer of fungus is seen with lots of browning and decaying, causing severe loss of leaves.

E. Mosaic Virus

Mosaic Virus in pulses is caused by several viral pathogens, including Bean Common Mosaic Virus (BCMV) and Cucumber Mosaic Virus (CMV). This disease causes mosaic patterns, leaf distortions, and stunted growth, resulting in major yield losses. The diseases often spread through insect vectors like aphids.

Initial Stage

Symptoms: Appearance of light green to yellow mosaic patterns on young leaves. Mild abnormalities can be seen in the leaves.

Plant Physiology: Viruses replicate within plant cells, causing a breakdown in chloroplast function.

Visual Indicators: Mild mosaic patterns and slight leaf curling (Gajbhiye & Dongre, 2013).

Intermediate Stage

Symptoms: Mosaic patterns become more prominent, and leaves show severe abnormalities and blistering. Growth is slowed, and the space between stems becomes shorter.

Plant Physiology: viral spread is extensive causing major problems in plant metabolism and growth.

Visual Indicators: Prominent mosaic patterns, severe leaf deformation, and blistering can be seen.

Advanced Stage

Symptoms: Plants that are severely affected display large mosaic patterns, and curled and deformed leaves. Pods may be deformed, and overall yield is significantly reduced.

Plant Physiology: A viral infection throughout the plant causes major metabolic and developmental issues.

Visual Indicators: Large mosaic patterns, severe leaf and pod deformation, and significant yield loss can be seen.

Table 2. Disease Symptoms

Disease	Fungal Images
Fusarium Wilt	
Ascochyta Blight	
Rust	
Powdery Mildew	

RESULT

Disease Prevalence and Impact

Our extensive investigation in the countryside spread of Fusarium wilt and Ascochyta blight in pulse crops are the dominant diseases with an approximate loss of up to 60% in some areas was the main finding. Fusarium wilt is caused by the soil-borne fungus Fusarium oxysporum, which attacks the vascular system of the plants, the very system of water and nutrient transportation. Wilting and yellowing of leaves occur due to this and, in the end, it is plant death, especially in those battle-like situations that have high humidity. Besides, Ascochyta blight, which is produced by fungal pathogens similar to Ascochyta rabiei and Ascochyta lentis, first appears as speaking at the leaves and shanks. The condition quite significantly influences the plant's ability to effectively photosynthesize, hence the significant reduction in seed quality and marketable yields (Belay et al., 2022). The socioeconomic aspect of these infections is quite severe, especially for farmers who work with limited land and depend on these crops for their livelihoods.

Development and Validation of the Diagnostic Tool

To find a solution to the problems spawned by these diseases, we came out with a diagnostic tool presented on the web which is based on a sound dataset consisting of images of both healthy and diseased pulse crops. This is made possible by a Convolutional Neural Network (CNN), that underwent detailed, heavy training and validates the data. The first tests of the validation exercise showed a very impressive score of over 90% in telling between healthy and unhealthy leaves. This accuracy capacity is essential in the sense that farmers require reliable plant assessments. The CNN, however, was the one that was primarily instructed to brandish pictures relevant to the major diseases plaguing pulse crops, hence users were able to have a natural on-time conversation with the AI on health problems facing the crops. Fungicide resistance is a rising problem causing a decrease in the usefulness of fungicides. The current diagnostics reveal that resistance to some fungicides is greatly reduced which might be the reason for the resurgence of plant diseases. The complex diagnostic capacity which the tool permits could substantially improve the farmers in arriving at timely decisions that will consequently be the measures for crops' minimum damage due to diseases.

Economic Impact Projections

Our economic study shows that employing this diagnostic instrument would lead to a growth in pulse crop per unit area harvested by as much as 30% in the affected regions. This potential boost in productivity is foreseen to be around 69.8% of farmer incomes, lifting many households above the poverty line. In the regions where pulse crops are staples, better yields will not only benefit individuals but also foster regional food security by producing more nutrient-rich food. Furthermore, there may be a decrease in conformable crops and an increase in the protein supply (Thakre et al., 2023).

Technical Performance and User Interface

We developed a web-based diagnostic tool, which is intended to be user-friendly and easily accessible to both farmers and agricultural practitioners who may not be technically savvy. Concurrently with that, the architecture of the CNN model was reworked so that it could process uploaded images in about 3 seconds without affecting the quality of the output. The quick feedback provided by the tool itself is of particular significance in the application of agriculture, where a judicious attempted intervention can decide the fate of an entire crop. Not only that, the system offers learning materials that demonstrate specific symptoms and give treatment courses for these diseases. The e-tool empowers farmers to effectively take steps forward from a point of ignorance to one where they have all the information they need, such as the identification and subsequent treatment of diseases. This then is a guide to the best ways to grow crops efficiently.

Recommendations for Future Research

The idea to make the diagnostic tool more effective, however, requires not just research but execution and several key initiatives need to be set for the future. Expansion of the tool's database so that it can handle the full range of diseases associated with pulse crops is the most straightforward way to diversify its application through data science approaches. Utilizing machine learning practices for perpetually reviving and updating the model will not only facilitate the tool to adapt to the changing environmental conditions and tremendously developing diseases but also ensure that the tool is capable of continuously improving itself. Furthermore, long-term studies should be carried out to analyze the tool's impact on agriculture sustainability and overall crop productivity. The provision of a technical tool for farming purposes is one of the demands for collaboration with agricultural extension (local) services. Collaboration with agricultural extension services would be a key resource to promote the tool in remote areas to farmers. The collaborations can also play the bridging role between technology and embarked agricultural practice, therefore speeding up the delivery of resources to the end-users for minute-to-minute successes (Belay et al., 2022).

Figure 2. Pulse Disease Detection Symptom

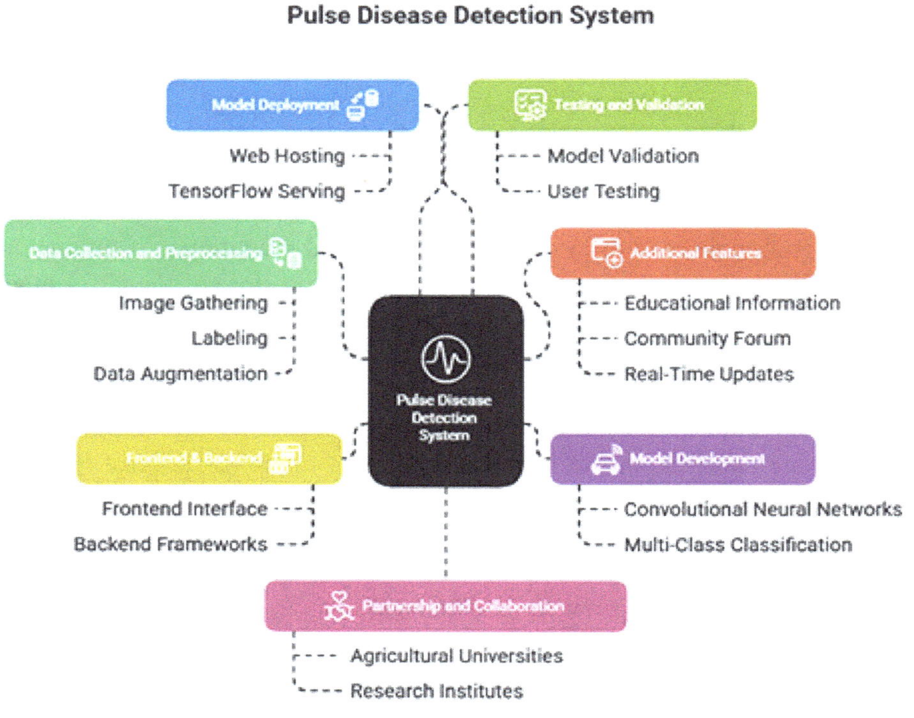

REFERENCES

Antichi, D., Pampana, S., Tramacere, L. G., Biarnes, V., Stute, I., Kadžiulienė, Ž., & Guilpart, N. (2023). An experimental dataset on yields of pulses across Europe. *Scientific Data, 10*(1), 708.

Belay, A. J., Salau, A. O., Ashagrie, M., & Haile, M. B. (2022). Development of a chickpea disease detection and classification model using deep learning. *Informatics in Medicine Unlocked, 31*, 100970.

Chen, Y. M., Chien, Y. Y., Chen, Y. K., Liao, P. Q., Tan, C. M., Chiu, Y. C., & Yang, J. Y. (2021). Identification of 16SrII-V phytoplasma associated with mungbean phyllody disease in Taiwan. *Plant Disease, 105*(9), 2290–2294.

Gajbhiye, K. G., & Dongre, S. S. (2013). A survey on weather monitoring system in agriculture zone using Zigbee. [IJSR]. *International Journal of Scientific Research, 2*(1).

Gautam, N. K., Kumar, K., & Prasad, M. (2016). Leaf crinkle disease in urdbean (Vigna mungo L. Hepper): An overview on causal agent, vector and host: NK Gautam et al. *Protoplasma, 253*(3), 729-746.

Giradkar, A., Adpawar, R., Agrawal, R., Dhule, C., & Morris, N. C. (2023, June). Design of Autonomous Weed Elimination using Maching Learning Techniques. In *2023 International Conference on Sustainable Computing and Smart Systems (ICSCSS)* (pp. 102-106). IEEE.

Gossen, B. D., Conner, R. L., Chang, K. F., Pasche, J. S., McLaren, D. L., Henriquez, M. A., & Hwang, S. F. (2016). Identifying and managing root rot of pulses on the northern great plains. *Plant Disease, 100*(10), 1965–1978.

Ingole, R. V., Nikhate, S. S., Agrawal, R., Dhule, C., & Morris, N. C. (2023, May). Enhancing plant health through deep neural network based leaf disease detection. In 2023 2nd International Conference on Applied Artificial Intelligence and Computing (ICAAIC) (pp. 1622-1627). IEEE.

Iqbal, J., Yousaf, U., Zia, S., Asgher, A., Afzal, R., Ali, M., & Sher, A. (2019). Pulses Diseases "Important limiting factor in yield"and their Managements. *Asian Journal of Research in Crop Science, 3*(2), 21.

Kalbande, K., Choudhary, S., Singru, A., Mukherjee, I., & Bakshi, P. (2021, June). Multi-way controlled feedback oriented smart system for agricultural application using internet of things. In 2021 5th International Conference on Trends in Electronics and Informatics (ICOEI) (pp. 96-101). IEEE.

Khaire, P. B., Hingole, D. G., Holkar, S. K., Ghante, P. H., & Mane, S. S. (2023). Management of Macrophomina phaseolina (Tassi.) Goid. causing leaf blight disease in mung bean in Maharashtra State. *Environment and Ecology*, *41*(2B), 1141–1148.

Martini, M., Delić, D., Liefting, L., & Montano, H. (2018). Phytoplasmas infecting vegetable, pulse and oil crops. In *Phytoplasmas: Plant Pathogenic Bacteria-I: Characterisation and Epidemiology of Phytoplasma-Associated Diseases* (pp. 31–65). Springer Singapore.

Pandhare, N., Mangde, M., Jajulwar, K., Mahule, M., & Patil, M. (2021, November). Design of multipurpose agro system using swarm intelligence. In *2021 International Conference on Computational Intelligence and Computing Applications (ICCICA)* (pp. 1-5). IEEE.

Pattanayak, A., Roy, S., Sood, S., Iangrai, B., Banerjee, A., Gupta, S., & Joshi, D. C. (2019). Rice bean: A lesser known pulse with well-recognized potential. *Planta*, *250*(3), 873–890.

Pillewan, M., Agrawal, R., Wyawahare, N., & Thakare, L. (2023, February). Review on design of smart domestic farming based on Internet of Things (IoT). In *2023 Third International Conference on Artificial Intelligence and Smart Energy (ICAIS)* (pp. 36-40). IEEE.

Rana, J. C., Gautam, N. K., Singh, M., Yadav, R., Tripathi, K., Yadav, S. K., & Bhardwaj, R. (2016). Genetic resources of pulse crops in India: An overview. *Indian Journal of Genetics and Plant Breeding*, *76*(04), 420–436.

Rashed, A., Feng, X., Prager, S. M., Porter, L. D., Knodel, J. J., Karasev, A., & Eigenbrode, S. D. (2018). Vector-borne viruses of pulse crops, with a particular emphasis on North American cropping system. *Annals of the Entomological Society of America*, *111*(4), 205–227.

Shende, D., Thakare, L., Agrawal, R., & Wyawahare, N. (2023, August). Automatic disease detection for various plants leaf using image processing techniques and tensorflow algorithm. In International Conference on Electrical and Electronics Engineering (pp. 487-499). Singapore: Springer Nature Singapore.

Shende, D., Wyawahare, N., Thakare, L., & Agrawal, R. (2023, February). Design process for adaptive spraying of pesticides based on mutual plant health detection and monitoring: a review. In *2023 Third International Conference on Artificial Intelligence and Smart Energy (ICAIS)* (pp. 729-733). IEEE.

Talasila, S., Rawal, K., Sethi, G., & Mss, S. (2022). Black gram Plant Leaf Disease (BPLD) dataset for recognition and classification of diseases using computer-vision algorithms. *Data in Brief*, *45*, 108725.

Tandekar, D., & Dongre, S. (2023, July). Identification of various diseases in plant leaves using image processing and CNN approach. In 2023 14th International conference on computing communication and networking technologies (ICCCNT) (pp. 1-6). IEEE.

Thakre, R. N., Kunte, P. A., Chavhan, N., Dhule, C., & Agrawal, R. (2023, November). UAV based system for detection in integrated insect management for agriculture using deep learning. In 2023 2nd International Conference on Futuristic Technologies (INCOFT) (pp. 1-6). IEEE.

Yasin, E. T., Kursun, R., & Koklu, M. (2023, August). Deep learning-based classification of black gram plant leaf diseases: A comparative study. In 11th International conference on advanced technologies (ICAT'23), Istanbul-Turkiye.

Zhang, C., Chen, T., Chen, W., & Sankaran, S. (2023). Non-invasive evaluation of Ascochyta blight disease severity in chickpea using field asymmetric ion mobility spectrometry and hyperspectral imaging techniques. *Crop Protection (Guildford, Surrey)*, *165*, 106163.

Chapter 9

A Comprehensive Analysis Comparing Deep Learning's Performance in Plant Leaf Identification and Disease Detection

Shagufta Shakeel
http://orcid.org/0009-0002-1820-272X
Madan Mohan Malaviya University of Technology, Gorakhpur, India

Sushil Kumar Saroj
Madan Mohan Malaviya, University of Technology, Gorakhpur, India

ABSTRACT

Agriculture is a significant element of the economy of a country, and the food that makes the nation healthier. Plant diseases should be determined to avoid wastage of resources and ensure the sustainability of the agricultural sector. Bacterial or animal diseases may seriously affect the food production system and the security of the global food supply. The application of deep learning has gained popularity because it has been used to diagnose and classify plant diseases within a very short period of time, on the basis of photos that are used.The study employs a hybrid deep learning model to predict the disease of bottle gourd and pumpkin leaves by the integration of EfficientNet, InceptionNet, and Transfer Learning. Our proposed models

DOI: 10.4018/979-8-3373-2429-6.ch009

performed better in terms of precision and strength over the existing methods, with an accuracy of 83% in predicting the illness of pumpkin leaf and 96% in predicting the illness of bottle gourd leaf on a real picture dataset with augmentation methods.

INTRODUCTION

By 2050, human civilisation is expected to improve food production by a criterion of 75% to feed the anticipated 9.7 billion people living in the world. The average loss due to infectious diseases and pests as of the moment is up to 38 per cent, and many farmers in developing countries have suffered up to 99% (Litskas et al., 2019) (TA1/4rko Ylu & Hanbay, 2019) (Li et al., 2021) (Dong et al., 2023). Yield losses caused by plant diseases have an economic impact on both domestic and global food production systems (Narayanan et al., 2022). Plant disease is a deviation from plant normal state that alters important processes or interrupts them. It significantly affects farm economics and production in agriculture and is mainly transmitted through bacteria, fungi, viruses, or microscopic animals (Shoaib et al., 2023). Nobody wants to farm or engage in agriculture due to the challenges that farmers are facing in the modern world on a daily basis. They are all moving to the modern cities to ensure that the future generation can live safely and be spared of the hurdles of the agriculture industry (Saleem et al., 2019). This organization, the Food and Agriculture organization of the United Nations, has reported that the transboundary plant diseases have proven a major threat greater than what it was a few decades ago due to intense agriculture, trade, globalization and the climatic change. What makes it even worse is that diseases are easier to spread around the globe than ever in the history. This is due to the fact that there are new diseases which can appear in the territory where they are still unknown and subsequently local knowledge to combat them is also unpresented. Higher threat of transboundary diseases also poses severe threat to the livelihoods of the vulnerable farmers on top of food and nutrition security of millions of people since such incidences may lead to loss of crops in large scale (OpenKnowledge, n.d.).

Leaves of a plant are the most significant part of the plant. When a leaf is infected, the life cycle of the plant will be disturbed. Diseases may affect the leaf, the stem, the fruit and other plant parts. The distribution of diseases occurs in different parts of the plant. Photosynthesis is possible only with the help of leaves, which could be said to be the most important element in the plant. The leaf diseases could directly affect the life cycle of a given plant at once. Plant diseases should therefore be identified at an early stage (Brahimi et al., 2017) (Rashid et al., 2021) (Mohanty et al., 2016) (Ramcharan et al., 2017) (Polder et al., 2019). Fig. 1 illustrates a six-step cycle of disease in plants. When the pathogen attacks a plant leaf initially, the pathogen grows within the cells, where it exhibits the symptoms. As the infection continues, the disease goes to the adjacent leaves, destroying the leaf tissue in increasing measure until the plant finally dies. The detection of plant diseases is considered to be an extremely vital problem. There are early and late blights. Early blight is a fungal disease, and late blight is a bacterial disease. The presence of plant diseases can be detected early enough, leading to more effective decisions as far as agricultural productivity is concerned. Affected plants often have signs or spots on the stems, fruits, leaves or blossoms. To be more exact, the anomalies can be detected with the help of the specific patterns left as the result of each of the disease and pest conditions. Detection of plant disease is time-consuming and skilful. A farmer can also save money and waste in case he/she can detect them in their initial stages and treat them accordingly. By 25 years, it is likely to have a population of more than 9 billion people in the world. Food output will have to rise by 70 per cent to match the growing population's food requirement. The issue of crop disease is highly rated in most countries, particularly agricultural countries. Moreover, upon determining the type of crop infection, a visual investigation is subjective, lengthy to carry out and must, in most instances, be inaccurate, especially in the diseases that are identified by experts or farmers. In the identification of the disease affecting the plant, this may cause the use of the wrong medication, which may contaminate the environment and compromise the quality of the crop. Technological developments have the capacity to revolutionise the lives of farmers by giving them access to several different automated technologies. With a basic digital camera, a farmer can quickly capture an image of his or her plant part and upload it to a disease detection system to learn about recommended pesticides and other remedies. The fact that deep learning is becoming a popular phenomenon is also of great importance. Very long, it will keep on making the daily life of people convenient. Plant disease affects the quality of the crop, the yield and the leaves, fruits, stems and roots. This has resulted in the lack of vegetables consumed globally. 16 percent of the agricultural production declines annually due to crop diseases. To address these limitations, the establishment of a hybrid deep learning architecture, which would fuse Transfer Learning, Inception-Net, and EfficientNet to produce a scalable and highly precise output in terms of identifying plant leaf diseases, is proposed in our work. To match the field to new tools, methods, and equipment, deep learning is a typical practice of smart farming. It deals with problematic topics in agriculture, including photo categorize diseases, should therefore be identified at an early stage. Fig. 1 illustrates a six step cycle in disease in plants. When the pathogen attacks a plant leaf initially the pathogen grows within the cells where they exhibit the symptoms. As the infection continues the disease goes to the adjacent leaves destroying the leaf tissue in increasing measure until the plant finally dies. The detection of plant diseases is considered to be an extremely vital problem. There are early and late blight. Early blight is the fungal disease and late blight is a bacterial disease specific. The presence of the plant diseases can be detected early enough leading to more effective decisions as far as the agricultural productivity is concerned. Affected plants often have signs or spots on the stems, fruits, leaves or blossoms. To be more exact, the anomalies can be detected with the help of the specific patterns left as the result of each of the disease and pest conditions. Detection of plant disease is time-consuming and skillful. A farmer can also save money and wastes in case he/she can detect them in their initial stages and treat them accordingly. By 25 years, it is likely to be having a population of more than 9 billion people in the world. Food output will have to rise by 70% to match the growing population food requirement. The issue with crop disease is highly rated in most countries particularly agricultural countries. Moreover, upon determining the type of crop infection, a visual investigation is subjective, lengthy to carry out and must in most instances be inaccurate, especially in the diseases that are identified by experts or farmers. In identification of the disease affecting the plant, this may cause the use of wrong medication that may contaminate the environment and compromise on the quality of crop.

Figure 1. Life Cycle of Plant Leaf Disease

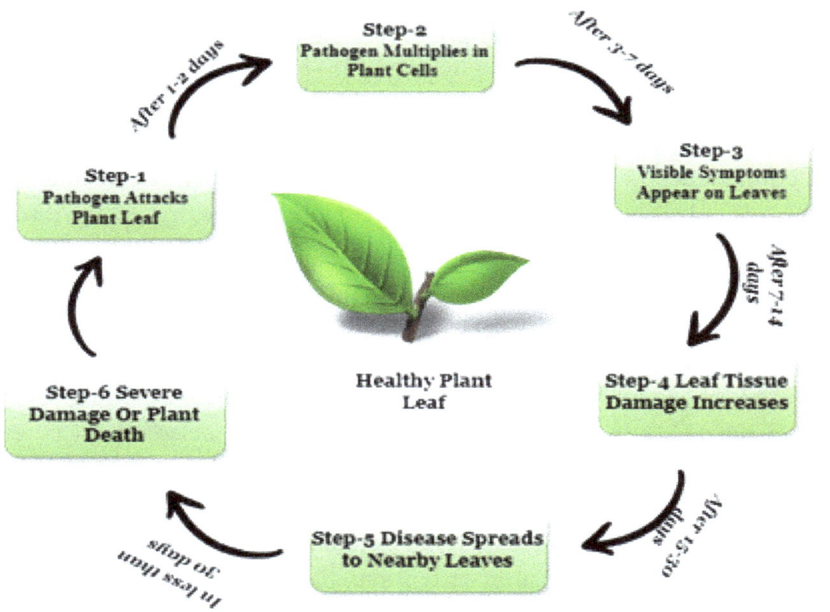

Technological developments have the capacity to revolutionize the lives of farmers by giving them access to several different automated technology (Jadhav et al., 2020)

(Jadhav & Rasal, 2024). With a basic digital camera, a farmer can quickly capture an image of his or her plant part and upload it to a disease detection system to learn about recommended pesticides and other remedies. The fact that deep learning is becoming a popular phenomenon is also of great importance. Very long, it will keep on making the daily life of people convenient. Plant disease affects the quality of the crop, the yield and the leaves, fruits, stems and the roots (Panchal et al., 2021). This has resulted in the lack of vegetables consumed globally. 16% of the agricultural production declines annually due to crop diseases. To address these limitations, the establishment of hybrid deep learning architecture, which would fuse Transfer Learning, InceptionNet, and EfficientNet to produce the scalable and highly precise output in terms of identifying plant leaf diseases, is proposed in our work. To match the field to new tools, methods, and equipment, deep learning is a typical practice of smart farming. It deals with problematic topics in agriculture including photo categorization, pattern recognition, feature extraction and transformation. The goal of the work is to create a CNN-based solution to identify crop diseases and the type of diseases. Consequently, the proposed model is more effective than the current one.

Motivation: This study aims to develop a hybrid, composite deep learning model which bypasses the limitations of single and traditional CNN models. Transition Learning allows the model to learn effectively using agricultural data sets that are scarce, but using separate neural nets like InceptionNet is known to be best at gathering multi-scale features, and EfficientNet is best at achieving a balance between depth and breadth to serve its purpose best. In our hybrid model, we have the complementary skills of EfficientNet (lightweight and scale-based architecture), Transfer Learning (flexible to work on smaller and specific data sets) and InceptionNet (path naming feature extraction). Though lacking adequately labelled data, the model is able to generalise to a great variety of plant diseases due to its aggregation. The algorithm aids farmers in diagnosing their crops early, applying pesticides in time, and reducing crop losses since it automatically identifies plant diseases based on the images of the leaves. It enhances food security systems, promotes the sustainability of agricultural practices, and there are financial benefits that arise since it reduces the dependence on the manual inspection process and the domain knowledge. By means of an effective hybrid modelling strategy, our research will eventually help to bridge the distance between the latest deep learning breakthroughs and mutually beneficial agricultural factor.

The rest of the paper are organized as follow: section 2 presents the literature review, section 3 presents the dataset overview, section 4 presents the proposed methodology, section 5 presents the results and discussion and section 6 presents the conclusion and future scope.

LITERATURE REVIEW

This section presents the literature and research on the use of deep learning in the detection of the disease in plant leaves and highlights the increased importance of this approach to diagnosing plant leaf diseases and preventing their further development. It discusses how picture classification and identification algorithms can be deceived, based on how CNNs rely on color feature extraction, and how such methods are essential to the diagnosis of sicknesses. Table 1 illustrates the overview of existing research in plant leaf detection.

Table 1. Related work of plant leaf disease detection

Author ?(tm)s Name	Technique Used	Contributions	Outcome
(Lee et al., 2020)	CNN	Using visual symptoms to identify disease without relying on species	High accuracy, and generalizable to various crops
(Ahmad et al., 2023)	Review paper	Gaps found include real-time problems and a lack of standard datasets.	Urged standardized DB and real-time deployment
(Jackulin & Murugavalli, 2022)	DL & ML	addressed the identification of human errors that are common in Indian farming.	Early diagnosis to enhance yield
(Islam et al., 2023)	ResNet-50	A technique for classifying leaf diseases in real time.	High accuracy, optimized for real-field use
(Kadam et al., 2024)	CNN + Web API + Docker	Potato leaf categorization in real time using a browser.	Achieved 98.44% accuracy
(Pai et al., 2024)	DL	DL based on CV for detecting rice illness.	Higher than traditional ML in accuracy & speed
(Lu et al., 2017)	CNN	Classification of wheat diseases using spatial awareness.	Fast diagnosis in real-time on field
(Nagasubramanian et al., 2018)	3D-CNN + Hyperspectral Imaging	Using spectral and spatial characteristics to diagnose soybean disease.	95.73% accuracy, outperformed 2D methods
(Praveen Kumar et al., 2022)	ML	Using visual analysis to anticipate diseases in their early stages.	Early diagnosis to reduce yield loss
(Zhang et al., 2018)	CNN + Transfer Learning	Detecting tomato disease with ResNet.	97.28% accuracy with SGD
(Gai et al., 2023)	Self-supervised Contrastive Learning	Identification of cherry ripeness without labeling.	Improved generalization in small samples
(Kos et al., 2025)	CNN	Maize leaf disease captured via a mobile device.	Real-time, low-cost diagnosis

continued on following page

Table 1. Continued

Author ?(tm)s Name	Technique Used	Contributions	Outcome
(Ahmed & Reddy, 2021)	CNN	14-crop field-based plant disease classifier.	94% accuracy using 96k+ images
(Andrew et al., 2022)	Pretrained CNN	Classification of plant diseases by transfer learning.	99.81% accuracy, early & accurate diagnosis
(Wang et al., 2024)	Self-Supervised Learning	21-disease general plant disease classifier.	99.61% Acc, 98.51% F1-score
(Pandian et al., 2022)	14-Layer DCNN	Large-scale illness detection (147,500 images).	Higher than transfer models, scalable
(Rado ?aj et al., 2024)	InceptionV3 + IncMB Block	Detecting tomato disease using multi-scale characteristics.	97.78% accuracy
(Khan et al., 2018)	Deep Feature + CCDF	Fruit disease using the fusion approach.	98.60% accuracy, high precision
(Sun et al., 2017)	26-layer DL model	Identifying ornamental plants from 10,000 images.	91.78% accuracy on wild images
(Wang et al., 2019)	Object Detection + CNN	Identification of tomato leaf disease.	Robust for remote smart agriculture use

DATASET OVERVIEW

A comprehensive description of the data that is used to train and evaluate the proposed hybrid model is listed in the section. This information contains the sources of the datasets, the constructions of the images, classes, their proportion, and their use in the tasks connected with classes of the disease of the plants and the solutions of this problem with the help of deep learning. In particular, there are two datasets in the present study, Pumpkin Leaf Disease, and Plant Leaf Freshness (PLF). The contribution of each of these datasets confirms the resilience and the flexibility of the hybrid model.

Plant Leaf Freshness

The Plant Leaf Freshness (PLF) dataset has remained the most popular since its initial publication to train and develop deep learning-based models to diagnose and predict the severity of plant diseases. Overall, there are 13 319 photos of healthy and unhealthy leaves of crops that are commonly grown in Bangladesh. There are high-quality images of six different vegetables contained in this dataset, including 2223 images of bitter gourds, 1803 images of bottle gourds, 2449 images of tomatoes, 2944 images of eggplants, 1598 images of cauliflowers and 2159 images of cucumbers. Automated disease detection systems of plant life may be easier to test and train due to the availability of well-structured and well labeled images in the PLF dataset that are isolated on a plain background. In this investigation, we relied on the original available publicly available PLF dataset to prove the effectiveness of our model. The data was obtained using an open-sources platform, Kaggle, which ensures its reusability and availability to scientific research. Table 2 represents the description of the dataset of plant leaf disease.

Table 2. Plant Leaf Freshness Dataset Description

Crop name	No. Of images	Leaf types
Bitter Gourd	2223	Fresh + Diseased
Bottle Gourd	1803	Fresh + Diseased
Tomato	2449	Fresh + Diseased
Eggplant	2944	Fresh + Diseased
Cauliflower	1598	Fresh + Diseased
Cucumber	2159	Fresh + Diseased

Figure 2. Bottle gourd leaf

Fig. 2a: Healthy leaf Fig. 2b : Mildly infected leaf Fig. 2c: Severely diseased leaf

Fig. 2: Bottle gourd leaf

In this study we have particularly looked at the Bottle Gourd class under PLF that contains samples of leaves in the following three conditions: health, mildly diseased and severely harmed. In Fig. 2, the visual difference can be seen in the form of dark spots, holes, yellow discoloration or wilting. The leaf of a green bottle gourd as shown in Fig. 2(a) has a clear structure of the veins. The leaf has a healthy vein pattern and is in a triangular shape, which shows that it is in good health and that it has no indication of stress or some disease. As seen in Fig. 2 (b), there is a green leaf that has evident wrinkles with brown spots and holes with dry edges. All these traits could be due to dehydration, environmental stress, pest loss, disease or bug infections. Even the heart shaped leaf in Fig. 2(c) has brown spots on the leaves, visible veins and a greenish-yellow color, which could be under disease stress, excess water, or nutrient deficiency symptoms. Heart-shaped leaves are seen on many plant species, including gourds and pumpkin. Table 3 represents the bottle gourd leaf stages, symptoms and conditions

Table 3. Bottle gourd leaf stages, symptoms, and conditions

Plant leaf freshness (Bottle gourd)	Bottle gourd leaf stages	Symptoms	Conditions
Fresh leaf		Smooth texture, vivid green hue, no discoloration or stains, and no wilting or curling symptoms.	ideal growth circumstances, which include enough water, sunshine, and nutrients. No indications of illness or vermin infestation. The plant is doing well and growing.
Downey mildew		Spots on the top surfaces of leaves that are yellow or pale green. Especially in damp situations, you could notice fluffy or downy growth that ranges from white to purple on the bottom. Affected leaves may prematurely drop, curl, or distort.	thrives in damp, chilly conditions. Wet leaves and high humidity might encourage the disease's spread. Overcrowding of plants and inadequate air circulation might make the issue worse.
Anthracnose		On leaves, there are tiny, asymmetrical brown or yellow dots that gradually get darker and larger. These patches might become sunken sores. Pinkish spore masses may grow in the middle of these areas during rainy conditions. Leaf withering and dropping can be a symptom of severe illnesses.	happens often in warm, humid weather. Splashing water, contaminated equipment, or contaminated seeds can all disseminate the illness. Prolonged leaf wetness and high humidity might exacerbate the illness.

Pumpkin Leaf Disease Datasets

The pool of Pumpkin Leaf Disease data including 2000 high-resolution images of pumpkin leaves is a huge addition to the current dataset of plant diseases. This dataset aims at supporting research and diagnosis of numerous afflictions to pumpkin plants and can also be used in differentiation of healthy and sick leaves. Strong deep learning-based models of sickness recognition can be formed by changing the specifications of the photo acquisition environment. Nevertheless, a several images in this series represent whole crops or a number of diseased leaves. The collection itself has been very thoroughly arranged in several directories, each one devoted to a particular group of pumpkin leaf types and conditions. This systematic methodology facilitates data management and increases its accessibility for a variety of applications and research requirements. Table 4 shows the description of the pumpkin leaf disease dataset.

Table 4. Pumpkin leaf disease datasets description

Class name	No. Of images	Leaf types
Healthy	400	Healthy
Bacterial leaf spot	400	Diseased
Downy mildew	400	Diseased
Powdery mildew	400	Diseased
Mosaic disease	400	Diseased

Figure 3. Pumpkin leaf

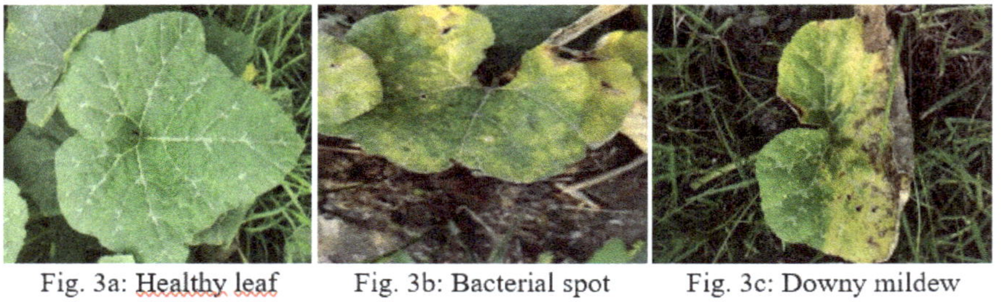

Fig. 3a: Healthy leaf Fig. 3b: Bacterial spot Fig. 3c: Downy mildew

Fig. 3: Pumpkin leaf

The pumpkin leaf is well displayed with a clear separation in Fig. 3. In Fig. 3(a) there can be visible white veins on the giant green squash and this shows that the vascular tissue is healthy. But the yellowing of the veins may have to mean that there is a nutritional deficiency or being over-watered and also it can be a disease. Grass and its plants are natural leaves. The presence of veins as shown in Fig. 3(b) in the leaf indicates that the leaf can transmit nutrients and water, and small defects on the rough surface of the leaf can be evidence of an early disease or insect infestation. Brown patches and pale green pigments on the yellow-green leaf shown in Fig. 3(c) could indicate disease or putrefaction. The jagged lines indicate invasions of insects or the collapse. The appearance of brown spots and white patches could indicate the presence of bacterial or fungal infection, nutrient deficiency or may be as a result of environmental stresses. The leaf can have more stressors since it is enclosed by grass. Table 5. Illustrates the pumpkin leaf stages, symptoms, and conditions

Table 5. Pumpkin leaf stages, symptoms, and conditions

Pumpkin Leaf Diseases Datasets	Pumpkin Leaf Stages	Symptoms	Conditions
Healthy Leaf		Large, heart-shaped, green, and rough-textured leaves are indicative of a healthy pumpkin. They feature noticeable veins, enough of surface area for photosynthesis, and pointy tips and lobed margins. Deviations might be a sign of an illness or a nutritional deficit.	ideal growing circumstances, which include enough sunshine, water, and nutrients. No indications of illness or vermin infestation.
Bacterial Leaf Spot		Tiny, light brown centers are frequently surrounded by yellow-brown dots on leaves.	Usually appears in warm, humid weather. Splashing water and watering above might disperse the microorganisms.
Downy Mildew		The upper surface of the leaves has yellow angular dots, while the underside has a downy, purplish-gray mold. In extreme circumstances, leaves may prematurely wither and perish.	thrives in damp, chilly environments. Wet leaves and high humidity might encourage the disease's spread.
Powdery Mildew		Powdery white or grayish patches on fruit, stems, and foliage. The plant can lose its leaves and yield fewer fruits.	usually takes place in warm, dry weather with high humidity. The issue may be made worse by crowded plants and inadequate air movement.
Mosaic Disease		On leaves with chlorotic halos, necrotic specks have the potential to develop into sizable, approximately circular lesions. Yellowing and dying leaves are a sign of severe infestations.	distributed by other insects, including aphids. can spread via contaminated instruments or seeds as well. It frequently happens in warm weather.

METHODOLOGY USED

In this study, a systematic and well-defined technique was used to efficiently detect and categorize plant leaf diseases. The method goes through several steps, from gathering data to utilizing deep learning models to forecast the disease in the end. Every stage of the pipeline, such as feature extraction, classification, augmentation, and image preprocessing, was created to guarantee the model's precision, resilience, and generalizability. To extract complex patterns from leaf photos, sophisticated neural network designs including CNNs, EfficientNet, and InceptionNet were used. Each step of the approach is thoroughly explained in this part, along with the reasoning behind the decisions made and how they affect the system's overall performance.

1. Dataset Collection and Preprocessing

The suggested hybrid model was tested on two distinct datasets of plant images, where the first one depicted the freshness index of bottle gourd leaves, whereas, the latter denoted the diseases of pumpkin leaves. No special preprocessing was done to each dataset including RGB images with labels organized in subdirectories per class. The images were standardized by scaling the pixel values to a range of 0-1 and all of these were resized to a standard input size (224 x 224 or 380 x 380, depending on the dataset). Folder names were encoded using one-hot and transformed into numeric classes to make labels. There was heavy data augment to make the model more general and robust. They included random flips, spins, zooming, variation of colour-spaces, alteration of brightness and contrast and displacement. In order to extend diversity training samples, the pumpkin set was augmented with a new CutMix augmentation method. In order to maintain the class distribution, stratified sampling was adopted to split the datasets into the training and validation sets.

2. Model Architecture Design

The proposed hybrid model ?(tm)s architecture is shown below.

Figure 4. Proposed model architecture

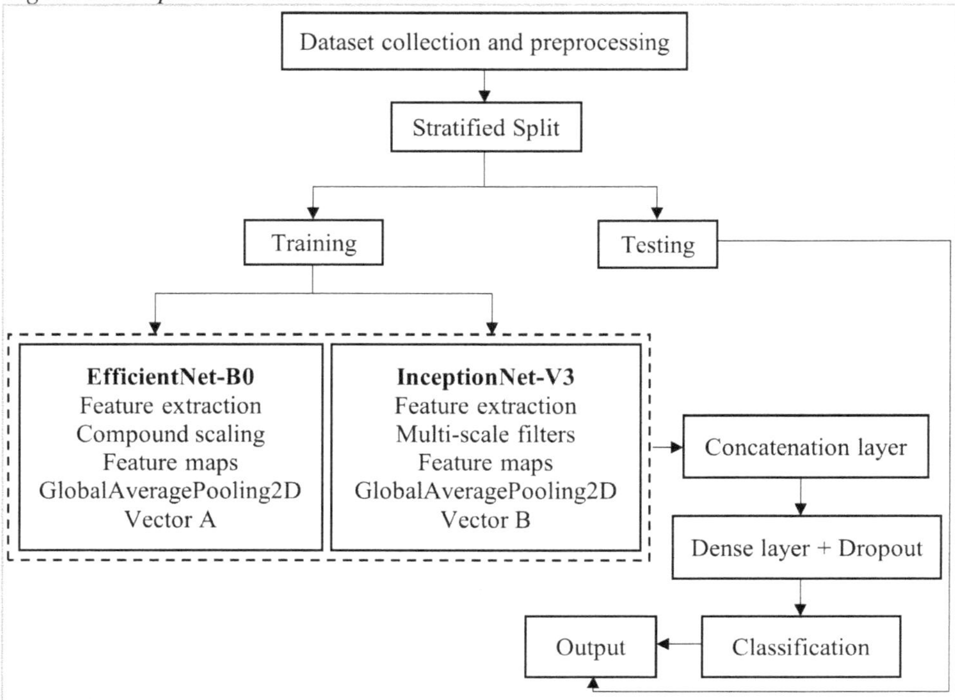

The suggested design is based on a hybrid convolutional neural network that uses EfficientNet and InceptionNet (InceptionV3) as a feature extractor in parallel as shown in Fig. 4. It was able to achieve convergence and better performance through faster convergence transfer of networks by initializing both with ImageNet pretrained weights. The highest layers of these models have been removed to add layers of categorization of our choice. Combined feature representation was extracted by concatenating feature maps that were extracted in EfficientNet and InceptionNet after they were separately passed through GlobalAveragePooling2D and GlobalAveragePooling2D layers independently. For the prevention of overfitting, this fused vector was fed into thick layers that were fully connected and used dropout regularization with ReLU activation functions. A softmax classifier was used as the final layer of the output, depending on the number of classifications that each of the datasets had. This architecture ensures fine-grained understanding of texture and high-level learning of the semantics (through the integration of the benefits of both of the CNN backbones). Fig. 4 demonstrates the architecture of the model.

3. Model Training

Model training was done in two phases. A training process kept regular EfficientNet and InceptionNet layers fixed but just trained custom dense layers using Adam optimizer and categorical cross-entropy loss with optional label smoothing. Subsequently, the model could keep the general-purpose features of the pretrained base and acquire dataset-specific ones. In the fine-tuning phase, SGD momentum was applied to restart the training with a significantly reduced learning rate after picking deeper layers on both base nets became not frozen. Learning rate schedulers like cosine annealing were utilized to avoid occurring too early convergence and make the learning rates change dynamically. The two phases involved satisfying the imbalance of classes by calculating the weight of classes. The size and number of batches and epochs were set depending on the complication of the dataset and its size.

4. Model Evaluation

After the training, numerous measures were applied to the model. Precision, recall, F1-score and accuracy were computed based on comparison of the prediction in the validation set against the ground-truth label. A classification report was drawn to capture performance per class. The confusion matrices were also generated to see the extent to which the model would discriminate between classments and identify any regular and consistent misclassifications. All the end assessments were done with the best model checkpoint where checkpoint is saved during training based on the validation accuracy. The ratio of accurately predicted samples to total samples is known as accuracy, and it indicates how accurate the model is overall. Accuracy by itself, however, could not be enough in multi-class classification situations like plant disease detection. As a result, recall and accuracy were also calculated. We may better comprehend how many projected illness cases were actually right by looking at precision, which shows the percentage of genuine positive predictions among all positive predictions generated by the model. Conversely, recall quantifies the model's capacity to identify every true positive case, or the number of authentic illness samples that were successfully located. The F1-score, which is a balanced metric particularly helpful for addressing class imbalance or overlapping illness symptoms, is the harmonic mean of accuracy and recall. The classification report from the model's predictions on the test dataset was used to compute these metrics for each illness category. We were able to make additional adjustments to the model's architecture or dataset quality by examining these scores, which allowed us to determine which plant illnesses the model was accurately identifying and which ones were frequently misclassified. Equations (i), (ii), (iii), and (iv) shows the formulae for calculating all the metrics.

$$\text{Accuracy} = \frac{\text{TP} + \text{TN}}{TP + FP + TN + FN}A \quad \text{(i)}$$

$$\text{Precision} = \frac{\text{TP}}{TP + FP}A \quad \text{(ii)}$$

$$\text{Recall} = \frac{\text{TP}}{TP + AFN}A \quad \text{(iii)}$$

$$\text{F1Ascore} = 2 * \text{A}\frac{Precision * Recall}{Precision + Recall}\text{A} \quad \text{(iv)}$$

5. Performance Visualization

By plotting training/validation accuracy/loss curves of both states (first and fine-tuning), one was able to gain a better insight into the learning behavior of the model. The stability of convergence and any possible overfitting or underfitting was assessed with the help of these graphs. Moreover, confusion matrix heatmaps were produced with Seaborn to illustrate the strong and weak points of several categorizations across groups. Classification reports that included comprehensive metrics including accuracy, recall, F1-score, and support for every illness category were produced to further examine the model's performance. Due to class imbalance or visual resemblance with other categories, these data helped determine which specific illness types were regularly correctly predicted and which ones need more study. To demonstrate the gains made possible by transfer learning, the learning curves and performance plots for the base model training and fine-tuning phases were also compared. It was shown that fine-tuning accelerated convergence in addition to improving the model's capacity for generalization. To prevent overfitting and preserve training effectiveness, early stopping and learning rate scheduling strategies were used.

RESULT AND DISCUSSION

Datasets of bottle gourd and pumpkin leaves were employed in order to test the effectiveness of the proposed hybrid model. First, confusion matrices of classified performance were generated as observed in Fig. 5. Most samples have been identified accurately and thus, the confusion matrix of the bottle gourd dataset presented in Fig. 5(a) is highly accurate and has good predictive power. Likewise, the robustness of the model to the variance in terms of the leaf types is confirmed using the confusion matrix of pumpkin leaf dataset shown in Fig. 5(b) indicating a balanced classification with a low number of error-predictions. A great number of graphs of the training and validation performance were obtained, as in Fig. 6, to analyze the learning impact of the model on the bottle gourd dataset even more thoroughly. Optimal convergence, yet not overfitting, is signified by the training and validation accuracy curve aligned in Fig. 6(a) increasing steadily with epochs and the loss curve also aligned in Fig. 6(b) decreasing steadily. The different plots of training and validation accuracy of the model as shown in Fig. 6(c) and Fig. 6(d) prove the stability of the model. The precision curves in Fig. 6(e) and Fig. 6(f) and the recall curves in Fig. 6(g) and Fig. 6(h) demonstrate the ability of the model to minimize false positives and properly identify real positives, respectively. The high consistency and reliability of the model is proven by the F1-score graphs in Fig. 6(i) and Fig. 6(j) that confirm a reasonable balance between accuracy and recall. The performance measures of the pumpkin leave dataset are shown in Fig. 7 and the accuracy plot in Fig. 7(a) shows that the model has a good predictive capability. Similarly, the good performance of the model to generalize well, regardless of the diversity in the types of leaves can further be attributed to the precision, recall, and F1-score exhibited on Figure 7(b), Fig. 7(c), and Fig. 7(d). The results indicate that the proposed hybrid architecture can be used to accommodate a wider scope of agricultural disease classification that it does not only learns effectively on a larger dataset like on bottle gourd but also generalizes well to other plants like the pumpkin.

Figure 5. Confusion matrix of bottle gourd and pumpkin leaf

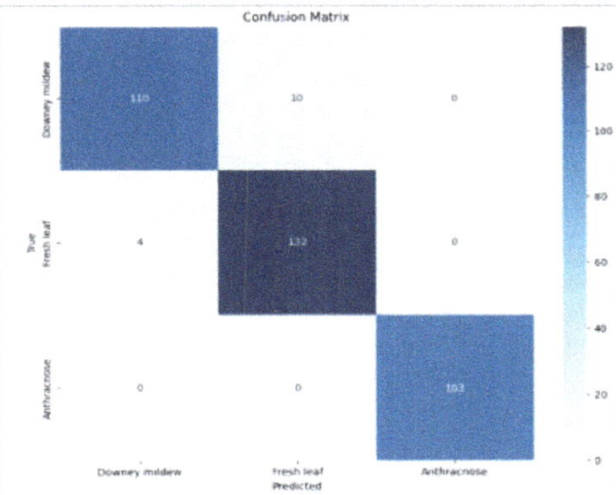

Fig. 5a : Confusion matrix of bottle gourd leaf

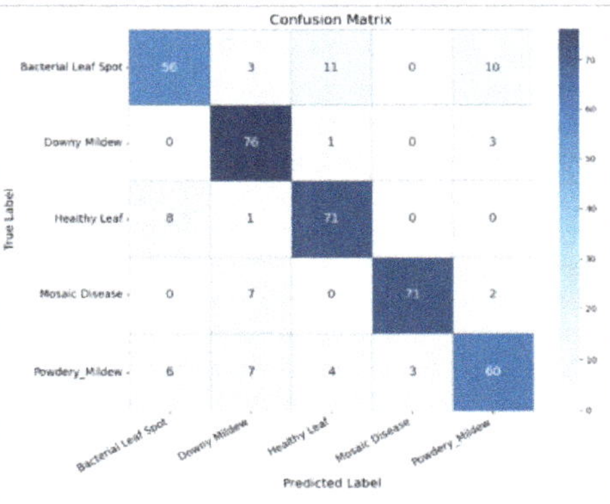

Fig. 5b : Confusion matrix of pumpkin leaf

Figure 6. Results of Bottle Gourd Leaf

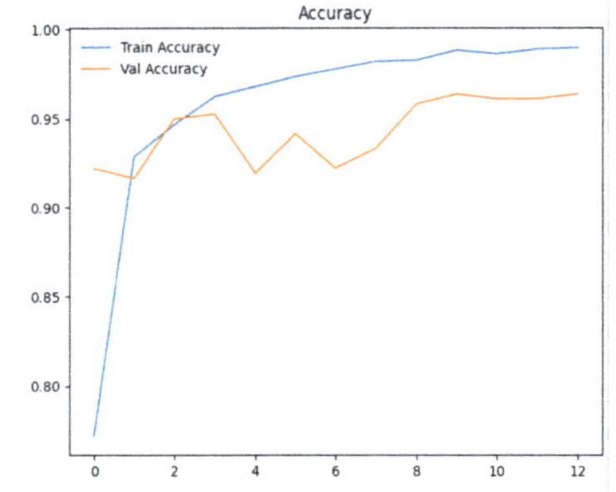

Fig.6(a : Training and validation accuracy

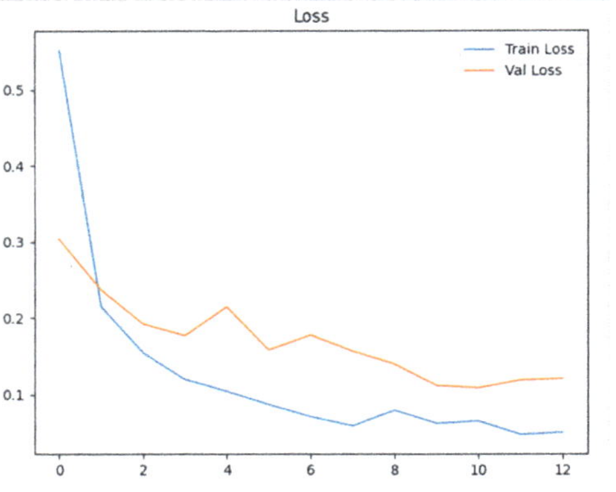

Fig.6(b) : Training and validation loss

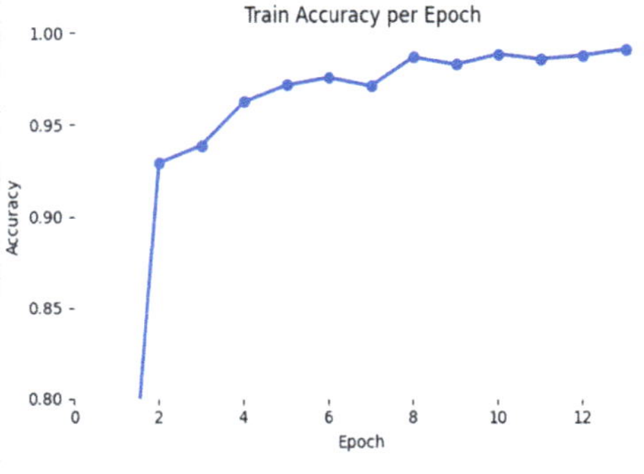

Fig.6(c) : Train accuracy

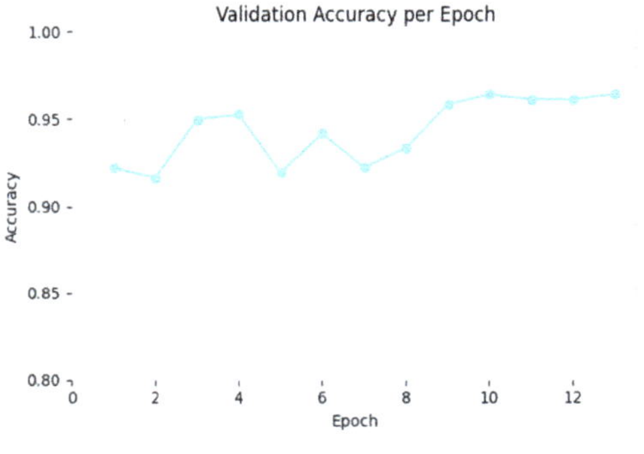

Fig.6(d) : Validation accuracy

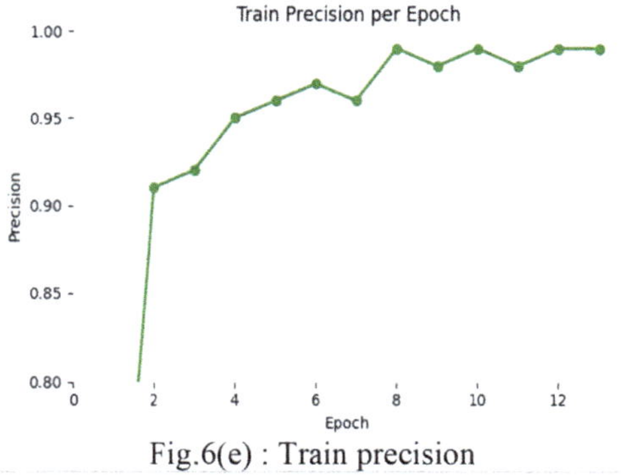

Fig.6(e) : Train precision

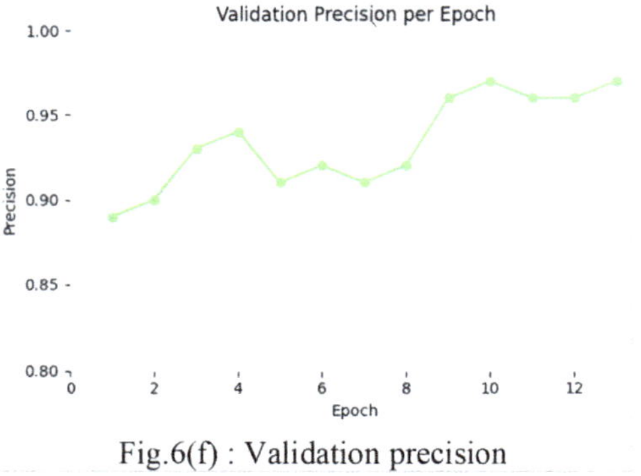

Fig.6(f) : Validation precision

Figure 7. Results of pumpkin leaf

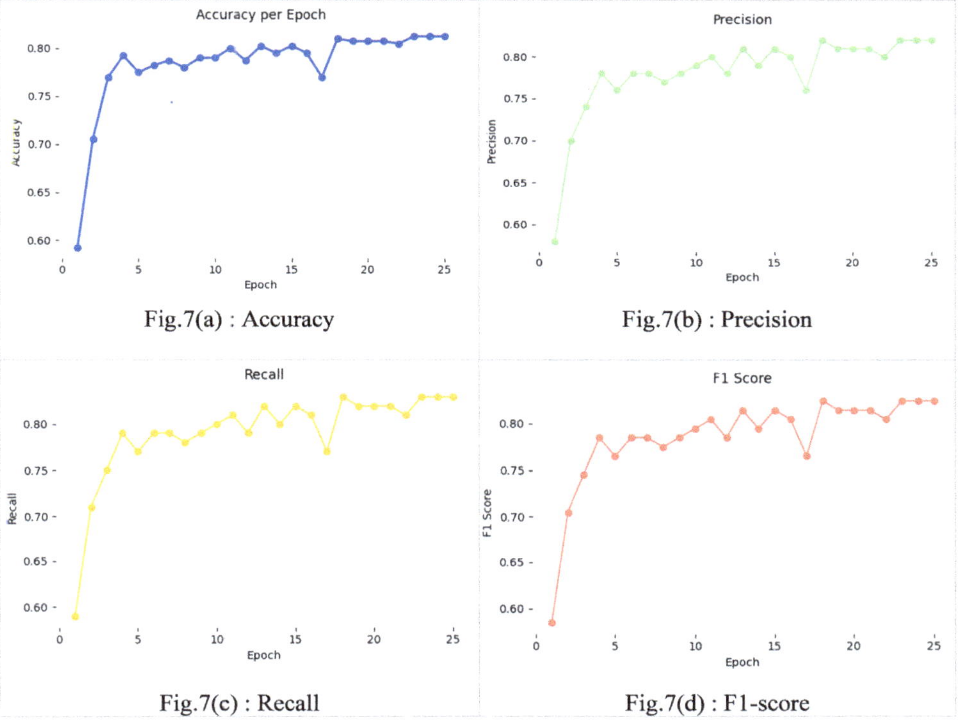

Fig.7(a) : Accuracy

Fig.7(b) : Precision

Fig.7(c) : Recall

Fig.7(d) : F1-score

In-depth analysis of performance at the class level was performed with classification reports, in addition to visual analysis on graphics.

Table 6. Classification report of bottle gourd leaf dataset

Class Label	Precision	Recall	F1-Score	Accuracy
Anthracnose	1.00	1.00	1.00	1.00
Downy Mildew	0.92	0.89	0.90	8.90
Fresh Leaf	0.97	1.00	0.98	1.00
Macro Average	0.96	0.96	0.96	-
Weighted Avg	0.96	0.96	0.96	-

Table 6 indicates the classification report of the bottle gourd leaf dataset. The bottle gourd dataset performed even better with the model having a macro F1-score of 0.96 as well as an overall accuracy of 96%. All three classes Downy Mildew, Fresh Leaf and Anthracnose, received high accuracy, recall and F1-scores and the last, Anthracnose, received a perfect 1.00 score on all measures. This proves the excellent generalizing ability of the model and its accuracy in categorizing the conditions of the bottle gourd leaves.

Table 7. Classification report of pumpkin leaf dataset

Class Label	Precision	Recall	F1-Score	Accuracy
Bacterial Leaf Spot	0.80	0.70	0.75	7.00
Downy Mildew	0.81	0.95	0.87	9.50
Healthy Leaf	0.82	0.89	0.85	8.90
Mosaic Disease	0.96	0.89	0.92	8.90
Powdery Mildew	0.80	0.75	0.77	7.50
Macro Average	0.84	0.83	0.83	-
Weighted Average	0.84	0.83	0.83	-

The overall accuracy of the model on the pumpkin leaf dataset was 83% with weighted F1-score of 0.83. As Table 7 shows, the class Mosaic Disease showed the best accuracy with F1-score of 0.96 and 0.92, revealing superior classification results of that disease. The relatively low recall of 0.70 and 0.75 F1-score accuracy on Bacterial Leaf Spot, however, indicates that there could be an individualized misclassification, due to the overlap of certain features to other categories of disease. The macro average F1-score of 0.83 further shows a balanced model performance across all the five classes in the pumpkin dataset.

CONCLUSION AND FUTURE SCOPE

The paper employed two datasets, i.e., pumpkin leaf disease and bottle gourd freshness, to build a hybrid deep learning model to detect and classify the diseases of plant leaves using EfficientNet and InceptionNet. The simultaneous extraction of semantic and fine-grained features followed by efficiently coupling features and classifying them using fully linked layers allowed the architecture to leverage best of both convolutional backbones. This was done by having high preprocessing which involved such processes as scaling, normalization, data augmentation such as using flipping, rotation, and using CutMix. Usage of transfer learning variant of weights pretrained on ImageNet was faster to converge and got better accuracy. The model had an encouraging performance in all the datasets proving its ability to accommodate different plant species and disease trends. It was also demonstrated that the model is reliable in the agricultural context through evaluation parameters such as accuracy,

F1-score etc. Although the bottle gourd dataset was more accurate, the performance of the pumpkin dataset is satisfactory but that also has improvement potential. The relatively low accuracy may be caused by a class imbalance or complex visual difference between illness categories. Augmentation with more aggressive methods, advanced regularization or balancing the classes such as oversampling or focus loss are all arbitrary techniques that have the potential to tremendously increase the accuracy of the pumpkin dataset. Future work may also include transformer-based visual models such as ViT or ConvNeXt or attention processes may also be used and center in an area of a leaf that is likely to be sick. Instant diagnosing may also benefit farmers in case the trained model is installed on mobile or edge devices so that field identification of plant diseases is possible in real time. Any addition of multimodal input, including soil-measurement or ambient-variable, may enhance the decision-making capacity of the model. On balance, the proposed mixture design offers a good foundation of intelligent, scalable, and useful plant disease monitoring systems in industrial agriculture.

REFERENCES

Ahmad, A., Saraswat, D., & El Gamal, A. (2023). A survey on using deep learning techniques for plant disease diagnosis and recommendations for development of appropriate tools. *Smart Agricultural Technology*, *3*, 2 ?"13. DOI: 10.1016/j.atech.2022.100083

Ahmed, A. A., & Reddy, G. H. (2021). A mobile-based system for detecting plant leaf diseases using deep learning. *AgriEngineering*, *3*(3), 478 ?"493. DOI: 10.3390/agriengineering3030032

Andrew, J., Eunice, J., Popescu, D. E., Chowdary, M. K., & Hemanth, J. (2022). Deep learning-based leaf disease detection in crops using images for agricultural applications. *Agronomy (Basel)*, *12*(10), 2 ?"19. DOI: 10.3390/agronomy12102395

Brahimi, M., Boukhalfa, K., & Moussaoui, A. (2017). Deep learning for tomato diseases: Classification and symptoms visualization. *Applied Artificial Intelligence*, *31*(4), 299 ?"315. DOI: 10.1080/08839514.2017.1315516

Dong, X., Wang, Q., Huang, Q., Ge, Q., Zhao, K., Wu, X., Wu, X., Lei, L., & Hao, G. (2023). PDDD-PreTrain: A Series of Commonly Used Pre-Trained Models Support Image-Based Plant Disease Diagnosis. *Plant Phenomics (Washington, D.C.)*, *5*, 1 ?"20. DOI: 10.34133/plantphenomics.0054 PMID: 37213546

Gai, R.-L., Wei, K., & Wang, P.-F. (2023). SSMDA: Self-supervised cherry maturity detection algorithm based on multi-feature contrastive learning. *Agriculture*, *13*(5), 2 ?"13. DOI: 10.3390/agriculture13050939

Islam, M. M., Adil, M. A. A., Ahamed, M. K. U., Rahman, M. M., Uddin, M. A., Talukder, M. A., Hasan, M. K., Sharmin, S., & Debnath, S. K. (2023). DeepCrop: Deep learning-based crop disease prediction with web application. *Journal of Agriculture and Food Research*, *14*, 2 ?"11. DOI: 10.1016/j.jafr.2023.100764

Jackulin, C., & Murugavalli, S. (2022). A comprehensive review on detection of plant disease using machine learning and deep learning approaches. *Measurement. Sensors*, *24*, 2 ?"10. DOI: 10.1016/j.measen.2022.100441

. Jadhav, S., & Rasal, R. (2024). ?oDeep learning based approach for plant disease detection and cure recommendation. ?? *International Journal of Research Publication and Reviews.*

Jadhav, S. B., Udupi, V. R., & Patil, S. B. (2020). Identification of Plant Diseases Using Convolutional Neural Networks. *International Journal of Information Technology : an Official Journal of Bharati Vidyapeeth ?(tm)s Institute of Computer Applications and Management*, *13*(6), 2461 ?"2470. DOI: 10.1007/s41870-020-00437-5

Kadam, P., Bidwe, R., Thorat, H., Shelke, S., Mahajan, K., Yewale, P., & Hiremani, V. (2024). Web-based application for plant leaf disease detection. *Journal of Electrical Systems.*, *20*(1s), 73 ?"83. DOI: 10.52783/jes.753

Khan, M. A., Akram, T., Sharif, M., Awais, M., Javed, K., Ali, H., & Saba, T. (2018). CCDF: Automatic System for Segmentation and Recognition of Fruit Crops Diseases Based on Correlation Coefficient and Deep CNN Features. *Computers and Electronics in Agriculture*, *155*, 220 ?"236. DOI: 10.1016/j.compag.2018.10.013

Kos, T., Zdrili , A., & Paja ? 1/2ivkovi , I. (2025). Towards smart pest management in olives: ANN-based detection of olive moth (*Prays oleae* Bernard, 1788). *AgriEngineering.*, *7*(7), 2 ?"12. DOI: 10.3390/agriengineering7070200

Lee, S. H., GoA"au, H., Bonnet, P., & Joly, A. (2020). New perspectives on plant disease characterization based on deep learning. *Computers and Electronics in Agriculture*, *170*, 2 ?"12. DOI: 10.1016/j.compag.2020.105220

Li, L., Zhang, S., & Wang, B. (2021). Plant Disease Detection and Classification by Deep Learning ?"A Review. *IEEE Access : Practical Innovations, Open Solutions*, *9*, 56683 ?"56698. DOI: 10.1109/ACCESS.2021.3069646

Litskas, V. D., Migeon, A., Navajas, M., Tixier, M.-S., & Stavrinides, M. C. (2019). Impacts of Climate Change on Tomato, a Notorious Pest and Its Natural Enemy: Small Scale Agriculture at Higher Risk. *Environmental Research Letters*, *14*(8), 2 ?"19. DOI: 10.1088/1748-9326/ab3313

Lu, J., Hu, J., Zhao, G., Mei, F., & Zhang, C. (2017). ?oAn in-field automatic wheat disease diagnosis system. ?? *arXiv*. pp. 2 ?"15. https://arxiv.org/abs/1710.08299

Mohanty, S. P., Hughes, D. P., & SalathA(c), M. (2016). Using Deep Learning for Image-Based Plant Disease Detection. *Frontiers in Plant Science*, *7*, 2 ?"10. DOI: 10.3389/fpls.2016.01419 PMID: 27713752

Nagasubramanian, K., Jones, S., Singh, A. K., Singh, A., Ganapathysubramanian, B., & Sarkar, S. (2018). ?oExplaining hyperspectral imaging based plant disease identification: 3D CNN and saliency maps. ?? pp. 2 ?"8. *arXiv*. https://arxiv.org/abs/1804.0883

Narayanan, K. L., Krishnan, R. S., Robinson, Y. H., Julie, E. G., Vimal, S., Saravanan, V., & Kaliappan, M. (2022). Banana Plant Disease Classification Using Hybrid Convolutional Neural Network. *Computational Intelligence and Neuroscience*, *2022*, 2 ?"13. DOI: 10.1155/2022/9153699 PMID: 35251158

OpenKnowledge. (n.d.). https://openknowledge.fao.org/server/api/core/bitstreams/30c0d98d-1c21-48ef-b5d9-8d988e6fa6f2/content

Pai, P., Amutha, S., Basthikodi, M., Gurpur, A. P., Chaitra, K. M., & Bhat, S. S. (2024). Deep learning-based classification methods for detection of diseases in rice leaves: A review. *International Journal of Intelligent Systems and Applications in Engineering.*, *12*(21s), 2064 ?"2077. DOI: 10.53555/ijisae.v12i21s.5775

Panchal, A. V., Patel, S. C., Bagyalakshmi, K., Kumar, P., Khan, I. R., & Soni, M. (2021). Image-Based Plant Diseases Detection Using Deep Learning. *Materials Today: Proceedings*. Advance online publication. DOI: 10.1016/j.matpr.2021.07.281

Pandian, J. A., Kumar, V. D., Geman, O., & Kanchanadevi, K. (2022). Plant disease detection using deep convolutional neural network. *Applied Sciences (Basel, Switzerland)*, *12*(14), 2 ?"17. DOI: 10.3390/app12146982

Polder, G., Blok, P. M., de Villiers, H. A. C., van der Wolf, J. M., & Kamp, J. (2019). Potato virus Y detection in seed potatoes using deep learning on hyperspectral images. *Frontiers in Plant Science*, *10*, 2 ?"13. DOI: 10.3389/fpls.2019.00209 PMID: 30881366

Praveen Kumar, K., Singh, B., Malini, K. V., Ranjan, A., Bonde, S. D., Rout, S., & Firos, A. (2022). *IoT and machine learning-based effective plant disease classification and detection for agricultural applications*. IJFANS International Journal of Food and Nutritional Sciences.

Rado ?aj, P., Rado ?aj, D., & Martinovi , G. (2024). Image-Based Leaf Disease Recognition Using Transfer Deep Learning with a Novel Versatile Optimization Module. *Big Data and Cognitive Computing*, *8*(6), 2 ?"14. DOI: 10.3390/bdcc8060052

Ramcharan, A., Baranowski, K., McCloskey, P., Ahmed, B., Legg, J., & Hughes, D. P. (2017). Deep Learning for Image-Based Cassava Disease Detection. *Frontiers in Plant Science*, *8*, 1852. Advance online publication. DOI: 10.3389/fpls.2017.01852 PMID: 29163582

. Rashid, J., Khan, I., Masood, K., Ali, G., Almotiri, S. H., & AlGhamdi, M. A. (2021). ?oMulti-level deep learning model for potato leaf disease recognition. ?? *Electronics. 10*(17). 2064. pp. 2-27. https://doi.org/.DOI: 10.3390/electronics10172064

Saleem, M. H., Potgieter, J., & Arif, K. M. (2019). Plant Disease Detection and Classification by Deep Learning. *Plants*, *8*(11), 2 ?"22. DOI: 10.3390/plants8110468 PMID: 31683734

Shoaib, M., Shah, B., El-Sappagh, S., Ali, A., Ullah, A., Alenezi, F., Gechev, T., Hussain, T., & Ali, F. (2023). An Advanced Deep Learning Models-Based Plant Disease Detection: A Review of Recent Research. *Frontiers in Plant Science*, *14*, 1 ?"22. DOI: 10.3389/fpls.2023.1158933 PMID: 37025141

Sun, Y., Liu, Y., Wang, G., & Zhang, H. (2017). Deep Learning for Plant Identification in Natural Environment. *Computational Intelligence and Neuroscience*, *2017*, 1 ?"6. DOI: 10.1155/2017/7361042 PMID: 28611840

TA1/4rko Ylu, M., & Hanbay, D. (2019). Plant Disease and Pest Detection Using Deep Learning-Based Features. *Turkish Journal of Electrical Engineering and Computer Sciences*, *27*(3), 1636 ?"1651. DOI: 10.3906/elk-1809-181

Wang, Q., Qi, F., Sun, M., Qu, J., & Xue, J. (2019). Identification of tomato disease types and detection of infected areas based on deep convolutional neural networks and object detection techniques. *Computational Intelligence and Neuroscience*, *2019*, 1 ?"15. DOI: 10.1155/2019/9142753 PMID: 31933623

Wang, Y., Yin, Y., Li, Y., Qu, T., Guo, Z., Peng, M., Jia, S., Wang, Q., Zhang, W., & Li, F. (2024). Classification of plant leaf disease recognition based on self-supervised learning. *Agronomy (Basel)*, *14*(3), 2 ?"13. DOI: 10.3390/agronomy14030500

Zhang, K., Wu, Q., Liu, A., & Meng, X. (2018). Can deep learning identify tomato leaf disease? *Advances in Multimedia*, *2018*, 1 ?"10. DOI: 10.1155/2018/6710865

Compilation of References

Abbas, M. S. T. (2018). Genetically engineered (modified) crops (Bacillus thuringiensis crops) and the world controversy on their safety. *Egyptian Journal of Biological Pest Control*, *28*(1), 1–12. DOI: 10.1186/s41938-018-0051-2

Abdollahi, J. (2022). Identification of medicinal plants in Ardabil using deep learning. In *2022 27th International Computer Conference, Computer Society of Iran (CSICC)* (pp. 1–6). IEEE.

Abdulridha, J., Ehsani, R., Abd-Elrahman, A., & Ampatzidis, Y. (2019). A remote sensing technique for detecting laurel wilt disease in avocado in the presence of other biotic and abiotic stresses. *Computers and Electronics in Agriculture*, *156*, 549–557. DOI: 10.1016/j.compag.2018.12.018

Abirami, A., Bhuvaneswari, S., Aadhithya, R., Devesh, R., & Deexith, P. (2023). *An empirical analysis on medicinal plants associated with impact on biodiversity, types, challenges, and technologies.*, DOI: 10.1109/ICACRS58579.2023.10404828

Acharya, L. K., Manik, S., Pradhan, M., Birah, A., Khokhar, M. K., Majumder, D., & Hossain, A. (2025). Recent Advancements in Genetic Engineering Techniques in Disease Resistance of Crops: An Outlook. In *Plant Stress Tolerance* (pp. 99-112). CRC Press.

Adimas, Z. T., Adimas, M. A., & Abera, B. D. (2024). Plant-based bioactive compounds for grain storage: a comprehensive review. In *Cogent Food and Agriculture* (Vol. 10, Issue 1). DOI: 10.1080/23311932.2024.2316152

Agrios, G. N. (2005). *Plant pathology* (5th ed.). Academic Press.

Ahmad, A., Saraswat, D., & El Gamal, A. (2023). A survey on using deep learning techniques for plant disease diagnosis and recommendations for development of appropriate tools. *Smart Agricultural Technology*, *3*, 2 ?"13. DOI: 10.1016/j.atech.2022.100083

Ahmar, S., Gill, R. A., Jung, K. H., Faheem, A., Qasim, M. U., Mubeen, M., & Zhou, W. (2020). Conventional and molecular techniques from simple breeding to speed breeding in crop plants: Recent advances and future outlook. *International Journal of Molecular Sciences*, *21*(7), 2590. DOI: 10.3390/ijms21072590 PMID: 32276445

Ahmed, A. A., & Reddy, G. H. (2021). A mobile-based system for detecting plant leaf diseases using deep learning. *AgriEngineering*, *3*(3), 478 ?"493. DOI: 10.3390/agriengineering3030032

Alae-Carew, C., Green, R., Stewart, C., Cook, B., Dangour, A. D., & Scheelbeek, P. F. D. (2022). The role of plant-based alternative foods in sustainable and healthy food systems: Consumption trends in the UK. *The Science of the Total Environment*, *807*, 151041. Advance online publication. DOI: 10.1016/j.scitotenv.2021.151041 PMID: 34673070

Ali, Q., Akhtar, T., Zuhra, N., Syarbiah, S., & Ega, A. A. (2024). Biotechnology for sustainable agriculture: Innovations in disease management. *Plant Prot*, *8*(4), 803–816. DOI: 10.33804/pp.008.04.5524

Ali, Q., Yu, C., Hussain, A., Ali, M., Ahmar, S., Sohail, M. A., Riaz, M., Ashraf, M. F., Abdalmegeed, D., Wang, X., Imran, M., Manghwar, H., & Zhou, L. (2022). Genome engineering technology for durable disease resistance: Recent progress and future outlooks for sustainable agriculture. *Frontiers in Plant Science*, *13*, 860281. DOI: 10.3389/fpls.2022.860281 PMID: 35371164

Ali, Z., Mahfouz, M. M., & Piatek, A. (2022). CRISPR/Cas genome editing: Precision tools for improving plant disease resistance. *Current Opinion in Biotechnology*, *75*, 102710.

Altieri, M. A., Nicholls, C. I., Henao, A., & Lana, M. A. (2015). Agroecology and the design of climate change-resilient farming systems. *Agronomy for Sustainable Development*, *35*(3), 869–890. DOI: 10.1007/s13593-015-0285-2

Altieri, M. A., Nicholls, C. I., & Montalba, R. (2017). Technological approaches to sustainable agriculture at a crossroads: An agroecological perspective. *Sustainability (Basel)*, *9*(3), 349. DOI: 10.3390/su9030349

Andrew, J., Eunice, J., Popescu, D. E., Chowdary, M. K., & Hemanth, J. (2022). Deep learning-based leaf disease detection in crops using images for agricultural applications. *Agronomy (Basel)*, *12*(10), 2 ?"19. DOI: 10.3390/agronomy12102395

Antichi, D., Pampana, S., Tramacere, L. G., Biarnes, V., Stute, I., Kadžiulienė, Ž., & Guilpart, N. (2023). An experimental dataset on yields of pulses across Europe. *Scientific Data*, *10*(1), 708.

Aroca-Fernandez, J. M., Diez-Pastor, J. F., Latorre-Carmona, P., Elvira, V., Camps-Valls, G., Pascual, R., & Garcia-Osorio, C. (2025). A Collaborative Platform for Soil Organic Carbon Inference Based on Spatiotemporal Remote Sensing Data. *arXiv preprint arXiv:2504.13962*.

Ashiq, A. (2024). Biopesticides in Sustainable Agriculture: Enhancing Targeted Pest Control and Ecosystem Health. *Applied Agriculture Sciences*, *2*(1), 1–8.

Azeem, A., Ul-Allah, S., Sher, A., Ijaz, M., Sattar, A., Ahmad, W., & Qayyum, A. (2023). Biosafety and biosecurity in genetically modified crops. In *Sustainable agriculture in the era of the OMICs revolution* (pp. 501–510). Springer International Publishing. DOI: 10.1007/978-3-031-15568-0_23

Babar, A. Z., & Akan, O. B. (2024). Sustainable and precision agriculture with the internet of everything (IoE). *arXiv preprint arXiv:2404.06341*.

Bacha, S. A. S., & Iqbal, B. (2023). Advancing agro-ecological sustainability through emerging genetic approaches in crop improvement for plants. *Functional & Integrative Genomics*, *23*(2), 145. DOI: 10.1007/s10142-023-01074-4 PMID: 37133756

Barimah, K. B., & Akotia, C. S. (2015). The promotion of traditional medicine as enactment of community psychology in Ghana. *Journal of Community Psychology*, *43*(1), 99–106. DOI: 10.1002/jcop.21687

Beaver, J. S., & Osorno, J. M. (2009). Achievements and limitations of contemporary common bean breeding using conventional and molecular approaches. *Euphytica*, *168*(2), 145–175. DOI: 10.1007/s10681-009-9911-x

Bebber, D. P., Ramotowski, M. A., & Gurr, S. J. (2013). Crop pests and pathogens move polewards in a warming world. *Nature Climate Change*, *3*(11), 985–988. DOI: 10.1038/nclimate1990

Bedő, Z., & Láng, L. (2015). Wheat breeding: current status and bottlenecks. In *Alien introgression in wheat: Cytogenetics, molecular biology, and genomics* (pp. 77–101). Springer International Publishing. DOI: 10.1007/978-3-319-23494-6_3

Belay, A. J., Salau, A. O., Ashagrie, M., & Haile, M. B. (2022). Development of a chickpea disease detection and classification model using deep learning. *Informatics in Medicine Unlocked*, *31*, 100970.

Bender, S. F., Wagg, C., & van der Heijden, M. G. (2016). An underground revolution: Biodiversity and soil ecological engineering for agricultural sustainability. *Trends in Ecology & Evolution*, *31*(6), 440–452. DOI: 10.1016/j.tree.2016.02.016 PMID: 26993667

Bradshaw, J. E. (2017). Review and analysis of limitations in ways to improve conventional potato breeding. *Potato Research*, *60*(2), 171–193. DOI: 10.1007/s11540-017-9346-z

Brahimi, M., Boukhalfa, K., & Moussaoui, A. (2017). Deep learning for tomato diseases: Classification and symptoms visualization. *Applied Artificial Intelligence*, *31*(4), 299 ?"315. DOI: 10.1080/08839514.2017.1315516

Bratspies, R. M. (2012). Is Anyone Regulating: The Curious State of GMO Governance in the United States. *Vermont Law Review*, *37*, 923.

Breiman, L. (2001). Random forests. *Machine Learning*, *45*(1), 5–32. DOI: 10.1023/A:1010933404324

Brooks, R. A. (1986). A robust layered control system for a mobile robot. *IEEE Journal on Robotics and Automation*, *RA-2*(1), 14–23. DOI: 10.1109/JRA.1986.1087032

Buziashvili, A., Kvitko, B., & Dangl, J. L. (2023). Engineering synthetic immunity and integrating ecological strategies for sustainable crop protection. *Trends in Plant Science*, *28*(5), 432–445.

Campbell, C. L., & Madden, L. V. (1990). *Introduction to plant disease epidemiology*. John Wiley & Sons.

Carney, J. A. (2021). Subsistence in the Plantationocene: Dooryard gardens, agrobiodiversity, and the subaltern economies of slavery. *The Journal of Peasant Studies*, *48*(5), 1075–1099. Advance online publication. DOI: 10.1080/03066150.2020.1725488

Carrière, Y., Brown, Z. S., Downes, S. J., Gujar, G., Epstein, G., Omoto, C., Storer, N. P., Mota-Sanchez, D., Søgaard Jørgensen, P., & Carroll, S. P. (2020). Governing evolution: A socioecological comparison of resistance management for insecticidal transgenic Bt crops among four countries. *Ambio*, *49*(1), 1–16. DOI: 10.1007/s13280-019-01167-0 PMID: 30903512

Ceccarelli, S., & Grando, S. (2007). Decentralized-participatory plant breeding: An example of demand driven research. *Euphytica*, *155*(3), 349–360. DOI: 10.1007/s10681-006-9336-8

Cellini, F., Chesson, A., Colquhoun, I., Constable, A., Davies, H. V., Engel, K. H., Gatehouse, A. M. R., Kärenlampi, S., Kok, E. J., Leguay, J.-J., Lehesranta, S., Noteborn, H. P. J. M., Pedersen, J., & Smith, M. (2004). Unintended effects and their detection in genetically modified crops. *Food and Chemical Toxicology*, *42*(7), 1089–1125. DOI: 10.1016/j.fct.2004.02.003 PMID: 15123383

Chakraborty, S., Pangga, I. B., Lupton, J., Hart, L., Room, P. M., & Yates, D. (2002). Production and dispersal of Colletotrichum gloeosporioides spores on Stylosanthes scabra under elevated CO_2. *Environmental Pollution*, *120*(3), 681–695. PMID: 15092933

Chaloner, T. M., Gurr, S. J., & Bebber, D. P. (2020). The global burden of plant disease tracks crop yields under climate change. bioRxiv, 2020-04.

Chaloner, T. M., Gurr, S. J., & Bebber, D. P. (2021). Plant pathogen infection risk tracks global crop yields under climate change. *Nature Climate Change*, *11*(8), 710–715. DOI: 10.1038/s41558-021-01104-8

Chanyal, H., Yadav, R. K., & Saini, D. K. J. (2022). Classification of medicinal plants leaves using deep learning technique: A review. *International Journal of Intelligent Systems and Applications in Engineering*, *10*(4), 78–87.

Chaudhuri, A., & Datta, A. (2018). Genetically modified (GM) crops: A potential source to combat global hunger and malnutrition. *Austin Journal of Nutrition and Food Sciences*, *6*(3), 1106.

Chen, Y. M., Chien, Y. Y., Chen, Y. K., Liao, P. Q., Tan, C. M., Chiu, Y. C., & Yang, J. Y. (2021). Identification of 16SrII-V phytoplasma associated with mungbean phyllody disease in Taiwan. *Plant Disease*, *105*(9), 2290–2294.

Chiti, T., Blasi, E., & Chiriacò, M. V. (2024). Carbon sequestration in a bamboo plantation: A case study in a Mediterranean area. *Journal of Forestry Research*, *35*(1), 51. Advance online publication. DOI: 10.1007/s11676-024-01696-9

Choudhary, B., Gheysen, G., Buysse, J., van der Meer, P., & Burssens, S. (2014). Regulatory options for genetically modified crops in India. *Plant Biotechnology Journal*, *12*(2), 135–146. DOI: 10.1111/pbi.12155 PMID: 24460889

Clapp, J., & Ruder, S. L. (2020). Precision technologies for agriculture: Digital farming, gene-edited crops, and the politics of sustainability. *Global Environmental Politics*, *20*(3), 49–69. DOI: 10.1162/glep_a_00566

Coakley, S. M., Scherm, H., & Chakraborty, S. (1999). Climate change and plant disease management. *Annual Review of Phytopathology*, *37*(1), 399–426. DOI: 10.1146/annurev.phyto.37.1.399 PMID: 11701829

Collard, B. C., & Mackill, D. J. (2008). Marker-assisted selection: An approach for precision plant breeding in the twenty-first century. *Philosophical Transactions of the Royal Society of London. Series B, Biological Sciences*, *363*(1491), 557–572. DOI: 10.1098/rstb.2007.2170 PMID: 17715053

Cook, D. E., Mesarich, C. H., & Thomma, B. P. (2015). Understanding plant immunity as a surveillance system to detect invasion. *Annual Review of Phytopathology*, *53*(1), 541–563. DOI: 10.1146/annurev-phyto-080614-120114 PMID: 26047564

Dale, J., James, A., Paul, Y., Khanna, H., Smith, M., Peraza-Echeverria, S., Garcia-Bastidas, F., Kema, G., Waterhouse, P., Mengersen, K., & Harding, R. (2017). Transgenic Cavendish bananas with resistance to Fusarium wilt tropical race 4. *Nature Communications*, *8*(1), 1496. DOI: 10.1038/s41467-017-01670-6 PMID: 29133817

Danso Ofori, A., Zheng, T., Titriku, J. K., Appiah, C., Xiang, X., Kandhro, A. G., Ahmed, M. I., & Zheng, A. (2025). The role of genetic resistance in rice disease management. *International Journal of Molecular Sciences*, *26*(3), 956. DOI: 10.3390/ijms26030956 PMID: 39940724

De Wolf, E. D., & Isard, S. A. (2007). Disease cycle approach to plant disease prediction. *Annual Review of Phytopathology*, *45*(1), 203–220. DOI: 10.1146/annurev.phyto.44.070505.143329 PMID: 17408356

Demilie, W. B. (2024). Plant disease detection and classification techniques: A comparative study of the performances. *Journal of Big Data*, *11*(1), 5. DOI: 10.1186/s40537-023-00863-9

Devkota, K., Ferreira, A. B., Timberlake, T. P., & dos Santos, C. F. (2024). The impact of pollinator decline on global protein production: Implications for livestock and plant-based products. *Global Ecology and Conservation*, *50*, e02815. Advance online publication. DOI: 10.1016/j.gecco.2024.e02815

Dey, S., Saha, S., Mandal, I., & Nandi, R. (2024). Sustainable Pest Management: As an Alternative to Chemical Pesticides. *International Journal of Scientific Research in Engineering and Management*, *8*(10), 1–5. DOI: 10.55041/IJSREM38087

Domingues, T., Brandão, T., & Ferreira, J. C. (2022). Machine learning for detection and prediction of crop diseases and pests: A comprehensive survey. *Agriculture*, *12*(9), 1350. DOI: 10.3390/agriculture12091350

Dong, X., & Ronald, P. C. (2019). Genetic engineering for disease resistance in plants: Recent progress and future perspectives. *Plant Physiology*, *180*(1), 26–38. DOI: 10.1104/pp.18.01224 PMID: 30867331

Dong, X., Wang, Q., Huang, Q., Ge, Q., Zhao, K., Wu, X., Wu, X., Lei, L., & Hao, G. (2023). PDDD-PreTrain: A Series of Commonly Used Pre-Trained Models Support Image-Based Plant Disease Diagnosis. *Plant Phenomics (Washington, D.C.)*, *5*, 1 ?"20. DOI: 10.34133/plantphenomics.0054 PMID: 37213546

EFSA Panel on Genetically Modified Organisms (GMO), Casacuberta, J., Barro, F., Braeuning, A., de Maagd, R., Epstein, M. M., ... & Dumont, A. F. (2025). Current practice, challenges and future opportunities in the safety assessment of newly expressed proteins in genetically modified plants. *EFSA Journal*, *23*(8), e9568.

Esquinas-Alcázar, J., Hilmi, A., & Noriega, I. L. (2012). A brief history of the negotiations on the International Treaty on Plant Genetic Resources for Food and Agriculture. In *Crop Genetic Resources as a Global Commons* (pp. 135–149). Routledge.

Ezrari, S., Legrifi, I., Taoussi, M., Khadiri, M., Belabess, Z., & Lahlali, R. (2024). Plant–pathogen interactions and global food security. In *Plant pathogen interaction* (pp. 11–52). Springer Nature Singapore.

Fahad, S., Saud, S., Akhter, A., Bajwa, A. A., Hassan, S., Battaglia, M., Adnan, M., Wahid, F., Datta, R., Babur, E., Danish, S., Zarei, T., & Irshad, I. (2021). Bio-based integrated pest management in rice: An agro-ecosystems friendly approach for agricultural sustainability. In *Journal of the Saudi Society of Agricultural Sciences* (Vol. 20, Issue 2). DOI: 10.1016/j.jssas.2020.12.004

Fenibo, E. O., Ijoma, G. N., & Matambo, T. (2022). Biopesticides in sustainable agriculture: Current status and future prospects. *New and future development in biopesticide research: Biotechnological exploration*, 1-53.

Ferentinos, K. P. (2018). Deep learning models for plant disease detection and diagnosis. *Computers and Electronics in Agriculture*, *145*, 311–318. DOI: 10.1016/j.compag.2018.01.009

Fernández Ríos, D., Quintana, S. A., Gómez Paniagua, P., Arrúa, A. A., Brozón, G. R., Bertoni Hicar, M. S., Castro Alegría, A., & Goberna, M. F. (2025). Regulatory challenges and global trade implications of genome editing in agriculture. *Frontiers in Bioengineering and Biotechnology*, *13*, 1609110. DOI: 10.3389/fbioe.2025.1609110 PMID: 40613005

Feuillet, C., & Eversole, K. (2023). An integrated, systems-wide approach is needed for public–private partnerships to drive genetic innovation in crops. *PLoS Biology*, *21*(7), e3002181. DOI: 10.1371/journal.pbio.3002181 PMID: 37410694

Flachs, A., Bastos, C., Heath, D., & Venkateswar, S. (2024). Introduction to Special Collection: Plant-Anthropo-Genesis: The Co-Production of Plant–People Lifeworlds. *Journal of Ethnobiology*, *44*(1), 3–10. Advance online publication. DOI: 10.1177/02780771241228068

Fontaine, S., Abbadie, L., Aubert, M., Barot, S., Bloor, J. M. G., Derrien, D., Duchene, O., Gross, N., Henneron, L., Le Roux, X., Loeuille, N., Michel, J., Recous, S., Wipf, D., & Alvarez, G. (2024). Plant–soil synchrony in nutrient cycles: Learning from ecosystems to design sustainable agrosystems. In *Global Change Biology* (Vol. 30, Issue 1). DOI: 10.1111/gcb.17034

Food and Agriculture Organization of the United Nations. (n.d.). *Agriculture on the proving grounds: Damage and loss*. Retrieved July 24, 2025, from https://www.fao.org/interactive/disasters-in-agriculture/en/

Food and Agriculture Organization of the United Nations. (n.d.). *Plant production and protection*. Retrieved September 1, 2025, from https://www.fao.org/plant-production-protection/about/

Food Security Information Network. (2025). *Global report on food crises 2025*. https://www.fsinplatform.org/report/global-report-food-crises-2025/

Francl, L. J. (2001). *The disease triangle: a plant pathological paradigm revisited*. The Plant Health Instructor., DOI: 10.1094/PHI-T-2001-0517-01

Fuchs, M., & Gonsalves, D. (2007). Safety of virus-resistant transgenic plants two decades after their introduction: Lessons from realistic field risk assessment studies. *Annual Review of Phytopathology*, *45*(1), 173–202. DOI: 10.1146/annurev.phyto.45.062806.094434 PMID: 17408355

Gai, R.-L., Wei, K., & Wang, P.-F. (2023). SSMDA: Self-supervised cherry maturity detection algorithm based on multi-feature contrastive learning. *Agriculture*, *13*(5), 2 ?"13. DOI: 10.3390/agriculture13050939

Gai, Y., & Wang, H. (2024). Plant disease: A growing threat to global food security. *Agronomy (Basel)*, *14*(8), 1615. DOI: 10.3390/agronomy14081615

Gajbhiye, K. G., & Dongre, S. S. (2013). A survey on weather monitoring system in agriculture zone using Zigbee. [IJSR]. *International Journal of Scientific Research*, *2*(1).

Gajjar, V. K., Nambisan, A. K., & Kosbar, K. L. (2022). Plant identification in a combined-imbalanced leaf dataset. *IEEE Access : Practical Innovations, Open Solutions, 10*, 37882–37891.

Gao, C. (2021). Genome engineering for crop improvement and future agriculture. *Cell, 184*(6), 1621–1635. DOI: 10.1016/j.cell.2021.01.005 PMID: 33581057

Gao, Y., Li, S. J., Zhang, S. W., Feng, T., Zhang, Z. Y., Luo, S. J., Mao, H.-Y., Borkovich, K. A., & Ouyang, S. Q. (2020). SlymiR482e-3p mediates tomato wilt disease by modulating ethylene response pathway. *Plant Biotechnology Journal, 19*(1), 17–19. DOI: 10.1111/pbi.13439 PMID: 32621307

Garrett, K. A., Dendy, S. P., Frank, E. E., Rouse, M. N., & Travers, S. E. (2006). Climate change effects on plant disease: Genomes to ecosystems. *Annual Review of Phytopathology, 44*(1), 489–509. DOI: 10.1146/annurev.phyto.44.070505.143420 PMID: 16722808

Gat, E. (1998). On three-layer architectures. In *Artificial intelligence and mobile robots* (Vol. 195, Article 210).

Gautam, N. K., Kumar, K., & Prasad, M. (2016). Leaf crinkle disease in urdbean (Vigna mungo L. Hepper): An overview on causal agent, vector and host: NK Gautam et al. *Protoplasma, 253*(3), 729-746.

Gay, P. B., & Gillespie, S. H. (2005). Antibiotic resistance markers in genetically modified plants: A risk to human health? *The Lancet. Infectious Diseases, 5*(10), 637–646. DOI: 10.1016/S1473-3099(05)70241-3 PMID: 16183518

Giradkar, A., Adpawar, R., Agrawal, R., Dhule, C., & Morris, N. C. (2023, June). Design of Autonomous Weed Elimination using Maching Learning Techniques. In *2023 International Conference on Sustainable Computing and Smart Systems (ICSCSS)* (pp. 102-106). IEEE.

Glick, B. R. (2014). Bacteria with ACC deaminase can promote plant growth and help to feed the world. *Microbiological Research, 169*(1), 30–39. DOI: 10.1016/j.micres.2013.09.009 PMID: 24095256

Golicz, A. A., Bayer, P. E., Bhalla, P. L., Batley, J., & Edwards, D. (2020). Pangenomics comes of age: From bacteria to plant and animal applications. *Trends in Genetics, 36*(2), 132–145. DOI: 10.1016/j.tig.2019.11.006 PMID: 31882191

Gossen, B. D., Conner, R. L., Chang, K. F., Pasche, J. S., McLaren, D. L., Henriquez, M. A., & Hwang, S. F. (2016). Identifying and managing root rot of pulses on the northern great plains. *Plant Disease, 100*(10), 1965–1978.

Grossman, M. R. (2019). Genetic engineering in the United States: regulation of crops and their food products. In *Regulation of genome editing in plant biotechnology: A comparative analysis of regulatory frameworks of selected countries and the EU* (pp. 263–312). Springer International Publishing. DOI: 10.1007/978-3-030-17119-3_7

Gu, J., Wang, Z., Kuen, J., Ma, L., Shahroudy, A., Shuai, B., Liu, T., Wang, X., Wang, G., Cai, J., & Chen, T. (2018). Recent advances in convolutional neural networks. *Pattern Recognition*, *77*, 354–377. DOI: 10.1016/j.patcog.2017.10.013

Guldenring, R., Andersen, R. E., & Nalpantidis, L. (2024). Zoom in on the Plant: Fine-Grained Analysis of Leaf, Stem, and Vein Instances. *IEEE Robotics and Automation Letters*, *9*(2), 1588–1595. Advance online publication. DOI: 10.1109/LRA.2023.3346807

Gumtow, R., Wu, D., Uchida, J., & Tian, M. (2018). A Phytophthora palmivora extracellular cystatin-like protease inhibitor targets papain to contribute to virulence on papaya. *Molecular Plant-Microbe Interactions*, *31*(3), 363–373. DOI: 10.1094/MPMI-06-17-0131-FI PMID: 29068239

Gupta, S., Kumar, A., Singh, P. K., & Joshi, A. K. (2023). Leaf rust in cereals: Challenges and solutions. *Microbial Biotechnology*, *18*(4), 567–580.

Han, K., Wang, Y., Chen, H., Chen, X., Guo, J., Liu, Z., Tang, Y., Xiao, A., Xu, C., Xu, Y., Yang, Z., Zhang, Y., & Tao, D. (2022). A survey on vision transformer. *IEEE Transactions on Pattern Analysis and Machine Intelligence*, *45*(1), 87–110. DOI: 10.1109/TPAMI.2022.3152247 PMID: 35180075

Harb, E. (2024, November). The Impact Of Public-Private Partnerships (Ppps) On Sustainable Agriculture And Food Security: A Comparative Analysis Of Different Models. In *ACE-International Conference* (pp. 298-301).

Harding, R., Paul, Y., James, A., Smith, M., Kleidon, J., Shekhawat, U., Phillips, A., Kidanemariam, D., Dawit, A., & Dale, J. (2025). QCAV-4, the first genetically modified Cavendish (cv. Grand Nain) banana resistant to Fusarium wilt tropical race 4 approved for commercial production and consumption. *Plant Biotechnology Journal*, *23*(9), 3628–3637. Advance online publication. DOI: 10.1111/pbi.70178 PMID: 40488601

Hassan, B. (2024). The Impact of Agricultural Policy on Global Food Security. *Frontiers in Agriculture*, *1*(2), 325–355.

Haverkort, A. J., Boonekamp, P. M., Hutten, R., Jacobsen, E., Lotz, L. A. P., Kessel, G. J. T., Vossen, J. H., & Visser, R. G. F. (2016). Durable late blight resistance in potato through dynamic varieties obtained by cisgenesis: Scientifc and societal advances in the DuRPh project. *Potato Research*, *59*(1), 35–66. DOI: 10.1007/s11540-015-9312-6

Henz, G. P. (2024). Impact of plant diseases on global food and nutritional security: A strategic theme for Brazil. *RAPP–Annual Review of Plant Pathology, Brasília*, *30*, 7–28. DOI: 10.31976/0104-038321v300001

Herring, R. J. (2014). On risk and regulation: Bt crops in India. *GM Crops and Food: Biotechnology in Agriculture and the Food Chain*, *5*(3), 204–209. DOI: 10.4161/21645698.2014.950543 PMID: 25437239

Hlaing, C. S., & Zaw, S. M. (2018). Tomato plant diseases classification using statistical texture feature and color feature. In *2018 IEEE/ACIS 17th International Conference on Computer and Information Science (ICIS)* (pp. 439–444). IEEE. DOI: 10.1109/ICIS.2018.8466483

Hochreiter, S., & Schmidhuber, J. (1997). Long short-term memory. *Neural Computation*, *9*(8), 1735–1780. DOI: 10.1162/neco.1997.9.8.1735 PMID: 9377276

Hossen, M. I., Awrangjeb, M., Pan, S., & Mamun, A. A. (2025). Transfer learning in agriculture: A review. *Artificial Intelligence Review*, *58*(4), 97. DOI: 10.1007/s10462-024-11081-x

Huang, K., Jahani, M., Gouzy, J., Legendre, A., Carrere, S., Lázaro-Guevara, J. M., González Segovia, E. G., Todesco, M., Mayjonade, B., Rodde, N., Cauet, S., Dufau, I., Staton, S. E., Pouilly, N., Boniface, M.-C., Tapy, C., Mangin, B., Duhnen, A., Gautier, V., & Rieseberg, L. H. (2023). The genomics of linkage drag in inbred lines of sunflower. *Proceedings of the National Academy of Sciences of the United States of America*, *120*(14), e2205783119. DOI: 10.1073/pnas.2205783119 PMID: 36972449

Huzenko, M., & Kononenko, S. (2024). Sustainable agriculture: Impact on public health and sustainable development. *Health Economics and Management Review*, *5*(2), 125–150. DOI: 10.61093/hem.2024.2-08

Ingole, R. V., Nikhate, S. S., Agrawal, R., Dhule, C., & Morris, N. C. (2023, May). Enhancing plant health through deep neural network based leaf disease detection. In 2023 2nd International Conference on Applied Artificial Intelligence and Computing (ICAAIC) (pp. 1622-1627). IEEE.

Iqbal, J., Yousaf, U., Zia, S., Asgher, A., Afzal, R., Ali, M., & Sher, A. (2019). Pulses Diseases "Important limiting factor in yield"and their Managements. *Asian Journal of Research in Crop Science*, *3*(2), 21.

Islam, M. M., Adil, M. A. A., Ahamed, M. K. U., Rahman, M. M., Uddin, M. A., Talukder, M. A., Hasan, M. K., Sharmin, S., & Debnath, S. K. (2023). DeepCrop: Deep learning-based crop disease prediction with web application. *Journal of Agriculture and Food Research*, *14*, 2 ?"11. DOI: 10.1016/j.jafr.2023.100764

Islam, T., Watanabe, K., De Silva, N. I., & Hyde, K. D. (2024). Resistance mechanisms of plant pathogenic fungi to fungicides, environmental impacts, and sustainable solutions. [MDPI]. *Plants*, *13*(19), 2737. DOI: 10.3390/plants13192737 PMID: 39409607

Ismail, B. I., Sehmi, M. N. M., Ahmad, H., Baharom, S. H., & Khalid, M. F. (2023). Robotic Research Platform for Agricultural Environment: Unmanned Ground Vehicle for Oil Palm Plantation. *Journal of Cases on Information Technology*, *25*(1), 1–32. Advance online publication. DOI: 10.4018/JCIT.328579

Jackulin, C., & Murugavalli, S. J. (2022). A comprehensive review on detection of plant disease using machine learning and deep learning approaches. *Measurement. Sensors*, *24*, 100441. DOI: 10.1016/j.measen.2022.100441

Jadhav, S. B., Udupi, V. R., & Patil, S. B. (2020). Identification of Plant Diseases Using Convolutional Neural Networks. *International Journal of Information Technology : an Official Journal of Bharati Vidyapeeth ?(tm)s Institute of Computer Applications and Management*, *13*(6), 2461 ?"2470. DOI: 10.1007/s41870-020-00437-5

Jafar, A., Bibi, N., Naqvi, R. A., Sadeghi-Niaraki, A., & Jeong, D. (2024). Revolutionizing agriculture with artificial intelligence: Plant disease detection methods, applications, and their limitations. *Frontiers in Plant Science*, *15*, 1356260. DOI: 10.3389/fpls.2024.1356260 PMID: 38545388

Javid, A., & Haghirosadat, B. F. (2017). A review of medicinal plants effective in the treatment or apoptosis of cancer cells. *Cancer Press Journal*, *3*(1), 22–26. DOI: 10.15562/tcp.41

Jayaraman, K. S. (2002). India approves GM cotton. *Nature Biotechnology*, *20*(5), 415–415. DOI: 10.1038/nbt0502-415 PMID: 11981538

Ji, W., Zhao, J., Liu, X., Deng, P., & Wang, C. (2025). Comparative transcriptomes reveal insights into different host responses associated with Fusarium head blight resistance in wheat. *BMC Plant Biology, 25*(509). Retrieved from BMC Plant Biology.

Jia, H., & Wang, N. (2020). Generation of homozygous canker-resistant citrus in the T0 generation using CRISPR-SpCas9p. *Plant Biotechnology Journal*, *18*(10), 1990–1992. DOI: 10.1111/pbi.13375 PMID: 32167662

Jones, J. D. G., & Dangl, J. L. (2006). The plant immune system. *Nature*, *444*(7117), 323–329. DOI: 10.1038/nature05286 PMID: 17108957

Jones, J. W., Antle, J. M., Basso, B., Boote, K. J., Conant, R. T., Foster, I., Godfray, H. C. J., Herrero, M., Howitt, R. E., Janssen, S., Keating, B. A., Munoz-Carpena, R., Porter, C. H., Rosenzweig, C., & Wheeler, T. R. (2017). Brief history of agricultural systems modeling. *Agricultural Systems*, *155*, 240–254. DOI: 10.1016/j.agsy.2016.05.014 PMID: 28701816

Kadam, P., Bidwe, R., Thorat, H., Shelke, S., Mahajan, K., Yewale, P., & Hiremani, V. (2024). Web-based application for plant leaf disease detection. *Journal of Electrical Systems.*, *20*(1s), 73 ?"83. DOI: 10.52783/jes.753

Kaelbling, L. P., Littman, M. L., & Moore, A. W. (1987). An architecture for intelligent reactive systems. In *Reasoning about actions and plans* (pp. 395–410). Elsevier. DOI: 10.1016/B978-0-934613-30-9.50019-6

Kalbande, K., Choudhary, S., Singru, A., Mukherjee, I., & Bakshi, P. (2021, June). Multi-way controlled feedback oriented smart system for agricultural application using internet of things. In 2021 5th International Conference on Trends in Electronics and Informatics (ICOEI) (pp. 96-101). IEEE.

Kamle, M., Borah, R., Bora, H., Jaiswal, A. K., Singh, R. K., & Kumar, P. (2020). Systemic acquired resistance (SAR) and induced systemic resistance (ISR): role and mechanism of action against phytopathogens. In *Fungal biotechnology and bioengineering* (pp. 457–470). Springer International Publishing. DOI: 10.1007/978-3-030-41870-0_20

Khaire, P. B., Hingole, D. G., Holkar, S. K., Ghante, P. H., & Mane, S. S. (2023). Management of Macrophomina phaseolina (Tassi.) Goid. causing leaf blight disease in mung bean in Maharashtra State. *Environment and Ecology*, *41*(2B), 1141–1148.

Khan, M. A., Akram, T., Sharif, M., Awais, M., Javed, K., Ali, H., & Saba, T. (2018). CCDF: Automatic System for Segmentation and Recognition of Fruit Crops Diseases Based on Correlation Coefficient and Deep CNN Features. *Computers and Electronics in Agriculture*, *155*, 220 ?"236. DOI: 10.1016/j.compag.2018.10.013

Kieu, N. P., Lenman, M., Wang, E. S., Petersen, B. L., & Andreasson, E. (2021). Mutations introduced in susceptibility genes through CRISPR/Cas9 genome editing confer increased late blight resistance in potatoes. *Scientific Reports*, *11*(1), 4487. DOI: 10.1038/s41598-021-83972-w PMID: 33627728

Kiflie, M., Sharma, D., Haile, M. A., & Srinivasagan, R. (2024). EfficientNet ensemble learning: Identifying Ethiopian medicinal plant species and traditional uses by integrating modern technology with ethnobotanical wisdom. *Computers*, *13*(2), 38. DOI: 10.3390/computers13020038

Kis, A., Hamar, É., Tholt, G., Bán, R., & Havelda, Z. (2019). Creating highly efficient resistance against wheat dwarf virus in barley by employing CRISPR/Cas9 system. *Plant Biotechnology Journal*, *17*(6), 1004–1006. DOI: 10.1111/pbi.13077 PMID: 30633425

Klindworth, D. L.. (2023). Linkage drag analysis in Aegilops speltoidesintrogressions for wheat improvement. [Springer]. *Theoretical and Applied Genetics*, *136*, 168. DOI: 10.1007/s00122-023-04409-8 PMID: 37410182

Kootstra, G., Bender, A., Perez, T., & van Henten, E. J. (2020). *Robotics in agriculture*. Springer., DOI: 10.1007/978-3-642-41610-1_43-1

Kos, T., Zdrili , A., & Paja ? 1/2ivkovi , I. (2025). Towards smart pest management in olives: ANN-based detection of olive moth (*Prays oleae* Bernard, 1788). *AgriEngineering.*, *7*(7), 2 ?"12. DOI: 10.3390/agriengineering7070200

Kovács, T., Botos, S., & Felföldi, J. (2024). What does digital transformation do in agriculture? A systematic literature review of Agri-food sector's digitalisation. *Agrárinformatika Folyóirat*, *14*(2). Advance online publication. DOI: 10.17700/jai.2023.14.2.707

Kranz, J. (1974). The role and scope of mathematical analysis and modeling in epidemiology. *Annual Review of Phytopathology*, *12*(1), 375–404. DOI: 10.1146/annurev.py.12.090174.002111

Kumar, A. S. (2024). The ethical and security implications of genetic engineering. *Observer Research Foundation Issue Brief, 723*

Kumar, A., Mahanta, D., Dange, M. M., Trivedi, A., & Nandeha, N. (2024). Global challenges facing plant pathology: A review on multidisciplinary approaches to meet the food security. *Journal of Scientific Research and Reports*, *30*(6), 884–892. DOI: 10.9734/jsrr/2024/v30i62106

Kuzma, J. (2018). Regulating gene-edited crops. *Issues in Science and Technology*, *35*(1), 80–85.

Lallawmkimi, M. C., Veda, D. S., Yadav, A., Kumar, M., & Rout, A. (2024). Innovative approaches in crop genetic engineering for sustainable agriculture: A review. *Journal of Advances in Biology & Biotechnology*, *27*(8), 615–631. DOI: 10.9734/jabb/2024/v27i81177

Lal, R. (2020). Regenerative agriculture for food and climate. *Journal of Soil and Water Conservation*, *75*(5), 123A–124A. DOI: 10.2489/jswc.2020.0620A

Lee, S. H., GoA"au, H., Bonnet, P., & Joly, A. (2020). New perspectives on plant disease characterization based on deep learning. *Computers and Electronics in Agriculture*, *170*, 2 ?"12. DOI: 10.1016/j.compag.2020.105220

Lee, S., Whitaker, V. M., & Hutton, S. F. (2016). Mini Review: Potential Applications of Non-host Resistance for Crop Improvement. *Frontiers in Plant Science*, *7*, 997. DOI: 10.3389/fpls.2016.00997 PMID: 27462329

Lemke, S. L. (2022). Gene Editing in Plants: A Nutrition Professional's Guide to the Science, Regulatory, and Social Considerations. *Nutrition Today*, *57*(2), 57–63. DOI: 10.1097/NT.0000000000000532

Li, J., Li, Y., & Ma, L. (2021). Recent advances in CRISPR/Cas9 and applications for wheat functional genomics and breeding. *aBIOTECH*, *2*(4), 375–385. DOI: 10.1007/s42994-021-00042-5 PMID: 36304421

Li, L., Zhang, S., & Wang, B. (2021). Plant disease detection and classification by deep learning—A review. *IEEE Access : Practical Innovations, Open Solutions*, *9*, 56683–56698. DOI: 10.1109/ACCESS.2021.3069646

Li, M. Y., Jiao, Y. T., Wang, Y. T., Zhang, N., Wang, B. B., Liu, R. Q., Yin, X., Xu, Y., & Liu, G. T. (2020). CRISPR/Cas9-mediated VvPR4b editing decreases downy mildew resistance in grapevine (*Vitis vinifera* L.). *Horticulture Research*, *7*(1), 7. DOI: 10.1038/s41438-020-00371-4 PMID: 32922821

Limenie, A. D., & Alehegn, M. (2025). Genetic engineering for cereal crop yield improvement and disease resistant breeding. *TheScientificWorldJournal*, *2025*(1), 6743917. DOI: 10.1155/tswj/6743917 PMID: 40225354

Li, T., Liu, B., Spalding, M. H., Weeks, D. P., & Yang, B. (2012). High-efficiency TALEN-based gene editing produces disease-resistant rice. *Nature Biotechnology*, *30*(5), 390–392. DOI: 10.1038/nbt.2199 PMID: 22565958

Litskas, V. D., Migeon, A., Navajas, M., Tixier, M.-S., & Stavrinides, M. C. (2019). Impacts of Climate Change on Tomato, a Notorious Pest and Its Natural Enemy: Small Scale Agriculture at Higher Risk. *Environmental Research Letters*, *14*(8), 2 ?"19. DOI: 10.1088/1748-9326/ab3313

Liu, L., Wang, J., Wang, F., & Yang, X. (2021). The impact of the planting of forest biomass energy plants under the embedded Internet of Things technology on the biodiversity of the local environmental ecology. *Environmental Technology & Innovation*, *24*, 101894. Advance online publication. DOI: 10.1016/j.eti.2021.101894

López-García, D., & Carrascosa-García, M. (2023). Agroecology-oriented farmers' groups. A missing level in the construction of agroecology-based local agri-food systems? *Agroecology and Sustainable Food Systems*, *47*(7), 996–1022. Advance online publication. DOI: 10.1080/21683565.2023.2217095

Lopez-Gomollon, S., & Baulcombe, D. C. (2022). Roles of RNA silencing in viral and non-viral plant immunity and in the crosstalk between disease resistance systems. *Nature Reviews. Molecular Cell Biology*, *23*(10), 645–662. DOI: 10.1038/s41580-022-00496-5 PMID: 35710830

Lovell-Read, F. A., Parnell, S., Cunniffe, N. J., & Thompson, R. N. (2023). Using 'sentinel' plants to improve early detection of invasive plant pathogens. *PLoS Computational Biology*, *19*(2), e1010884. Advance online publication. DOI: 10.1371/journal.pcbi.1010884 PMID: 36730434

Lu, J., Hu, J., Zhao, G., Mei, F., & Zhang, C. (2017). ?oAn in-field automatic wheat disease diagnosis system. ?? *arXiv*. pp. 2 ?"15. https://arxiv.org/abs/1710.08299

Lynch, R. E., Guo, B., Timper, P., & Wilson, J. P. (2003). United States Department of Agriculture–Agricultural Research Service research on improving host-plant resistance to pests. *Pest Management Science: Formerly. Pesticide Science*, *59*(6-7), 718–727. DOI: 10.1002/ps.636 PMID: 12846322

Mackelprang, R., & Lemaux, P. G. (2020). Genetic engineering and editing of plants: An analysis of new and persisting questions. *Annual Review of Plant Biology*, *71*(1), 659–687. DOI: 10.1146/annurev-arplant-081519-035916 PMID: 32023090

Macovei, A., Sevilla, N. R., Cantos, C., Jonson, G. B., Slamet-Loedin, I., Čermák, T., Voytas, D. F., Choi, I.-R., & Chadha-Mohanty, P. (2018). Novel alleles of rice eIF4G generated by CRISPR/Cas9-targeted mutagenesis confer resistance to Rice tungro spherical virus. *Plant Biotechnology Journal*, *16*(11), 1918–1927. DOI: 10.1111/pbi.12927 PMID: 29604159

Madden, L. V., Hughes, G., & Van Den Bosch, F. (2007). *The study of plant disease epidemics*. American Phytopathological Society Press.

Ma, H., Liu, Y., Zhang, S., Sha, J., Sun, Y., Hu, Z., & Ma, H. (2025). Wheat resistance to Fusarium head blight and breeding strategies. *Crop Health*, *3*(1), 1–11. DOI: 10.1007/s44297-025-00048-1 PMID: 41649652

Mahanayak, B. (2024). Ensuring biosafety of genetically modified organisms (GMOs): Regulatory frameworks and risk assessments in India. *International Journal of Research Publication and Reviews*, *5*(7), 1368–1373.

Mahfouze, H. A., & El-Sayed, O. E. (2022). Comparative analysis of late blight resistance R genes and their coding proteins in some potato genotypes. *Czech Journal of Genetics and Plant Breeding*, *58*(1), 10–20. DOI: 10.17221/49/2021-CJGPB

Mahlein, A. K. (2016). Plant disease detection by imaging sensors—Parallels and specific demands for precision agriculture and plant phenotyping. *Plant Disease*, *100*(2), 241–251. DOI: 10.1094/PDIS-03-15-0340-FE PMID: 30694129

Malik, S. A. (2024). SDG 2: Zero hunger, food and plant-related intellectual property, and access to plant genetic resources. In *The Elgar Companion to Intellectual Property and the Sustainable Development Goals*. DOI: 10.4337/9781803925233.00009

Malik, O. A., Ismail, N., Hussein, B. R., & Yahya, U. (2022). Automated realtime identification of medicinal plants species in natural environment using deep learning models—A case study from Borneo Region. *Plants*, *11*(15), 1952. DOI: 10.3390/plants11151952 PMID: 35956431

Mamabolo, E., Mashala, M. J., Mugari, E., Mogale, T. E., Mathebula, N., Mabitsela, K., & Ayisi, K. K. (2025). Application of precision agriculture technologies for crop protection and soil health. *Smart Agricultural Technology*, *12*, 101270. DOI: 10.1016/j.atech.2025.101270

Manoharan, J. S. (2021). Flawless detection of herbal plant leaf by machine learning classifier through two stage authentication procedure. *Journal of Artificial Intelligence and Capsule Networks*, *3*(2), 125–139. DOI: 10.36548/jaicn.2021.2.005

Martinelli, F., Scalenghe, R., Davino, S., Panno, S., Scuderi, G., Ruisi, P., Villa, P., Stroppiana, D., Boschetti, M., Goulart, L. R., Davis, C. E., & Dandekar, A. M. (2015). Advanced methods of plant disease detection: A review. *Agronomy for Sustainable Development*, *35*(1), 1–25. DOI: 10.1007/s13593-014-0246-1

Martini, M., Delić, D., Liefting, L., & Montano, H. (2018). Phytoplasmas infecting vegetable, pulse and oil crops. In *Phytoplasmas: Plant Pathogenic Bacteria-I: Characterisation and Epidemiology of Phytoplasma-Associated Diseases* (pp. 31–65). Springer Singapore.

Martin, T., Hossard, C., Therond, O., Baret, F., Colbach, N., Messéan, A., & Munier-Jolain, N. (2022). Robots and transformations of work in farm: A systematic review of the literature and a research agenda. *Agronomy for Sustainable Development*, *42*(4), 66. Advance online publication. DOI: 10.1007/s13593-022-00796-2

Mbabazi, R., Koch, M., Maredia, K., & Guenthner, J. (2021). Crop biotechnology and product stewardship. *GM Crops and Food: Biotechnology in Agriculture and the Food Chain*, *12*(1), 106–114. DOI: 10.1080/21645698.2020.1822133 PMID: 33079624

Mehra, A. (2020). *Agricultural robots market worth USD 20.3 billion by 2025*. MarketsandMarkets. https://www.marketsandmarkets.com/MarketReports/agricultural-robot-market-173601759.html

Mehta, D., Stürchler, A., Anjanappa, R. B., Zaidi, S. S. E. A., Hirsch-Hoffmann, M., Gruissem, W., & Vanderschuren, H. (2019). Linking CRISPR-Cas9 interference in cassava to the evolution of editing-resistant geminiviruses. *Genome Biology*, *20*(1), 80. DOI: 10.1186/s13059-019-1678-3 PMID: 31018865

Mills, W. D. (1944). Efficient use of sulfur dusts and sprays during rain to control apple scab. *Cornell Extension Bulletin*, *630*, 4.

Mohanty, S. P., Hughes, D. P., & Salathé, M. (2016). Using deep learning for image-based plant disease detection. *Frontiers in Plant Science*, *7*, 1419. DOI: 10.3389/fpls.2016.01419 PMID: 27713752

Mrisho, L. M., Mbilinyi, N. A., Ndalahwa, M., Ramcharan, A. M., Kehs, A. K., McCloskey, P. C., Murithi, H., Hughes, D. P., & Legg, J. P. (2020). Accuracy of a smartphone-based object detection model, PlantVillage Nuru, in identifying the foliar symptoms of the viral diseases of cassava–CMD and CBSD. *Frontiers in Plant Science*, *11*, 590889. DOI: 10.3389/fpls.2020.590889 PMID: 33391304

Murphy, R. R. (2019). *Introduction to AI robotics*. MIT Press.

Nagasubramanian, K., Jones, S., Singh, A. K., Singh, A., Ganapathysubramanian, B., & Sarkar, S. (2018). ?oExplaining hyperspectral imaging based plant disease identification: 3D CNN and saliency maps. ?? pp. 2 ?"8. *arXiv*. https://arxiv.org/abs/1804.0883

Narayanan, K. L., Krishnan, R. S., Robinson, Y. H., Julie, E. G., Vimal, S., Saravanan, V., & Kaliappan, M. (2022). Banana Plant Disease Classification Using Hybrid Convolutional Neural Network. *Computational Intelligence and Neuroscience*, *2022*, 2 ?"13. DOI: 10.1155/2022/9153699 PMID: 35251158

National Academies of Sciences, Medicine, Committee on Genetically Engineered Crops, Past Experience, & Future Prospects. (2016). Genetically engineered crops: experiences and prospects.

National Aeronautics and Space Administration. (n.d.). *Extreme weather and climate change*. Retrieved September 1, 2025, from https://science.nasa.gov/climate-change/extreme-weather/

National Research Council. (1993). *Managing Global Genetic Resources*. National Academies Press.

Niroui, F., Zhang, K., Kashino, Z., & Nejat, G. (2019). Deep reinforcement learning robot for search and rescue applications: Exploration in unknown cluttered environments. *IEEE Robotics and Automation Letters*, *4*(2), 610–617. DOI: 10.1109/LRA.2019.2891991

Nkhabindze, B. Z., Magagula, C. N., Earnshaw, D., Mhlanga, C. F., Matsebula, S. N., & Dladla, I. G. (2024). Regulatory framework for genetically modified organisms in the Kingdom of Eswatini. *GM Crops and Food: Biotechnology in Agriculture and the Food Chain*, *15*(1), 212–221. DOI: 10.1080/21645698.2024.2375664 PMID: 38963885

Noureen, A., Khan, M. Z., Amin, I., Zainab, T., & Mansoor, S. (2022). CRISPR/Cas9-mediated targeting of susceptibility factor eIF4E-enhanced resistance against Potato Virus Y. [Retrieved from Frontiers.]. *Frontiers in Genetics*, •••, 13.

Oerke, E. C. (2006). Crop losses to pests. *Journal of Agricultural Science*, *144*(1), 31–43. DOI: 10.1017/S0021859605005708

Ojiewo, C. O., Kugbei, S., Bishaw, Z., & Rubyogo, J. C. (2015). *Community Seed Production (Workshop Proceedings)* (pp. 1-176). ICRISAT.

OpenKnowledge. (n.d.). https://openknowledge.fao.org/server/api/core/bitstreams/30c0d98d-1c21-48ef-b5d9-8d988e6fa6f2/content

Orounladji, B. M., Sib, O., Berre, D., Assouma, M. H., Dabire, D., Sanogo, S., & Vall, E. (2024). Cross-examination of agroecology and viability in agro-sylvo-pastoral systems in Western Burkina Faso. *Agroecology and Sustainable Food Systems*, *48*(4), 581–609. Advance online publication. DOI: 10.1080/21683565.2024.2307902

Padma, T. V. (2022). When will India approve its first GM food crop? *Nature*, *611*(7937), 648–649. DOI: 10.1038/d41586-022-03738-w PMID: 36385294

Padol, P. B., & Yadav, A. A. (2016). SVM classifier based grape leaf disease detection. In *2016 Conference on Advances in Signal Processing (CASP)* (pp. 175–179). IEEE. DOI: 10.1109/CASP.2016.7746160

Pai, P., Amutha, S., Basthikodi, M., Gurpur, A. P., Chaitra, K. M., & Bhat, S. S. (2024). Deep learning-based classification methods for detection of diseases in rice leaves: A review. *International Journal of Intelligent Systems and Applications in Engineering.*, *12*(21s), 2064 ?"2077. DOI: 10.53555/ijisae.v12i21s.5775

Panchal, A. V., Patel, S. C., Bagyalakshmi, K., Kumar, P., Khan, I. R., & Soni, M. (2021). Image-Based Plant Diseases Detection Using Deep Learning. *Materials Today: Proceedings*. Advance online publication. DOI: 10.1016/j.matpr.2021.07.281

Pandey, K., Dangi, R., Prajapati, U., Kumar, S., Maurya, N. K., Singh, A. V., & Rajan, R. (2019). Advance breeding and biotechnological approaches for crop improvement: A review. *International Journal of Chemical Studies*, *7*(1), 837–841.

Pandhare, N., Mangde, M., Jajulwar, K., Mahule, M., & Patil, M. (2021, November). Design of multipurpose agro system using swarm intelligence. In *2021 International Conference on Computational Intelligence and Computing Applications (ICCICA)* (pp. 1-5). IEEE.

Pandian, J. A., Kumar, V. D., Geman, O., & Kanchanadevi, K. (2022). Plant disease detection using deep convolutional neural network. *Applied Sciences (Basel, Switzerland)*, *12*(14), 2 ?"17. DOI: 10.3390/app12146982

Pan, S. (2024). *Bt crops – Definition, types, advantages, limitations. Biology Notes Online*. Retrieved from Biology Notes Online.

Pattanayak, A., Roy, S., Sood, S., Iangrai, B., Banerjee, A., Gupta, S., & Joshi, D. C. (2019). Rice bean: A lesser known pulse with well-recognized potential. *Planta*, *250*(3), 873–890.

Paul, P. A., Lipps, P. E., Hershman, D. E., McMullen, M. P., Draper, M. A., & Madden, L. V. (2008). Efficacy of triazole-based fungicides for Fusarium head blight and deoxynivalenol control in wheat: A multivariate meta-analysis. *Phytopathology*, *98*(9), 999–1011. DOI: 10.1094/PHYTO-98-9-0999 PMID: 18943738

Peeriga, R., Balagani, P. K., & Atmakuri, L. R. (2022). Prominence of Biopesticides to supersede chemical pesticides in sustainable agriculture. *International Journal of Pharmacy and Pharmaceutical Sciences*, *5*(1), 81–85.

Perry, E. D., Ciliberto, F., Hennessy, D. A., & Moschini, G. (2016). Genetically engineered crops and pesticide use in US maize and soybeans. *Science Advances*, *2*(8), e1600850. DOI: 10.1126/sciadv.1600850 PMID: 27652335

Phani, V., Khan, M. R., & Dutta, T. K. (2021). Plant-parasitic nematodes as a potential threat to protected agriculture: Current status and management options. In *Crop Protection* (Vol. 144). DOI: 10.1016/j.cropro.2021.105573

Pillewan, M., Agrawal, R., Wyawahare, N., & Thakare, L. (2023, February). Review on design of smart domestic farming based on Internet of Things (IoT). In *2023 Third International Conference on Artificial Intelligence and Smart Energy (ICAIS)* (pp. 36-40). IEEE.

Pixley, K. V., Falck-Zepeda, J. B., Giller, K. E., Glenna, L. L., Gould, F., Mallory-Smith, C. A., Stelly, D. M., & Stewart, C. N.Jr. (2019). Genome editing, gene drives, and synthetic biology: Will they contribute to disease-resistant crops, and who will benefit? *Annual Review of Phytopathology*, *57*(1), 165–188. DOI: 10.1146/annurev-phyto-080417-045954 PMID: 31150590

Ploetz, R. C. (2015). Fusarium wilt of banana. *Phytopathology*, *105*(12), 1512–1521. DOI: 10.1094/PHYTO-04-15-0101-RVW PMID: 26057187

Polder, G., Blok, P. M., de Villiers, H. A. C., van der Wolf, J. M., & Kamp, J. (2019). Potato virus Y detection in seed potatoes using deep learning on hyperspectral images. *Frontiers in Plant Science*, *10*, 2 ?"13. DOI: 10.3389/fpls.2019.00209 PMID: 30881366

Popescu, D., Dragana, C., Stoican, F., Ichim, L., & Stamatescu, G. (2018). A collaborative UAV-WSN network for monitoring large areas. *Sensors (Basel)*, *18*(12), 4202. DOI: 10.3390/s18124202 PMID: 30513655

Praveen Kumar, K., Singh, B., Malini, K. V., Ranjan, A., Bonde, S. D., Rout, S., & Firos, A. (2022). *IoT and machine learning-based effective plant disease classification and detection for agricultural applications.* IJFANS International Journal of Food and Nutritional Sciences.

Pretty, J., Benton, T. G., Bharucha, Z. P., Dicks, L. V., Flora, C. B., Godfray, H. C. J., Goulson, D., Hartley, S., Lampkin, N., Morris, C., Pierzynski, G., Prasad, P. V. V., Reganold, J., Rockström, J., Smith, P., Thorne, P., & Wratten, S. (2018). Global assessment of agricultural system redesign for sustainable intensification. *Nature Sustainability, 1*(8), 441–446. DOI: 10.1038/s41893-018-0114-0

Prohens, J., Fita, A. M., Rodríguez-Burruezo, A., Raigón, M. D., Plazas, M., & Vilanova, S. (2011). Breeding for the Present and the Future: Achievements and Constraints of Conventional Plant Breeding and Contributions of Genomics to a New Green Revolution. *Bulletin of University of Agricultural Sciences and Veterinary Medicine Cluj-Napoca. Horticulture, 68*(1).

Qaim, M. (2003). Bt cotton in India: Field trial results and economic projections. *World Development, 31*(12), 2115–2127. DOI: 10.1016/j.worlddev.2003.04.005

Qaim, M. (2020). Role of new plant breeding technologies for food security and sustainable agricultural development. *Applied Economic Perspectives and Policy, 42*(2), 129–150. DOI: 10.1002/aepp.13044

Qin, G., He, W., Sanders, C. J., Zhang, J., Zhou, J., Wu, J., Lu, Z., Yu, M., Li, Y., Li, Y., Lambers, H., Li, Z., & Wang, F. (2024). Contributions of plant- and microbial-derived residuals to mangrove soil carbon stocks: Implications for blue carbon sequestration. *Functional Ecology, 38*(3), 573–585. Advance online publication. DOI: 10.1111/1365-2435.14497

Rado ?aj, P., Rado ?aj, D., & Martinovi , G. (2024). Image-Based Leaf Disease Recognition Using Transfer Deep Learning with a Novel Versatile Optimization Module. *Big Data and Cognitive Computing, 8*(6), 2 ?"14. DOI: 10.3390/bdcc8060052

Raj, M., Sharma, S., Gupta, A., & Nair, R. M. (2025). Mosaic Virus in Solanaceous Crops: Genetic Resistance and Biocontrol Approaches. *Tropical Plant Pathology, 30*(2), 78–95.

Ramasamy, S. P., Rathnasamy, S. A., Manickam, S., Chellappan, G., Natarajan, B., Swaminathan, M., Seshadri, G., & Muthurajan, R. (2025). Genome editing for improving disease resistance in rice (Oryza sativa L.)-Scope, achievements, limitations and future needs. *Physiological and Molecular Plant Pathology, 138*, 102640. DOI: 10.1016/j.pmpp.2025.102640

Ramcharan, A., Baranowski, K., McCloskey, P., Ahmed, B., Legg, J., & Hughes, D. P. (2017). Deep Learning for Image-Based Cassava Disease Detection. *Frontiers in Plant Science, 8*, 1852. Advance online publication. DOI: 10.3389/fpls.2017.01852 PMID: 29163582

Rana, J. C., Gautam, N. K., Singh, M., Yadav, R., Tripathi, K., Yadav, S. K., & Bhardwaj, R. (2016). Genetic resources of pulse crops in India: An overview. *Indian Journal of Genetics and Plant Breeding*, *76*(04), 420–436.

Rani, L., Devika, G., Karegowda, A. G., Vidya, S., & Bhat, S. (2022, April). Identification of medicinal leaves using state of art deep learning techniques. In *2022 IEEE International Conference on Distributed Computing and Electrical Circuits and Electronics (ICDCECE)* (pp. 1-5). IEEE.[

Rao, R. U., Lahari, M. S., Sri, K. P., Srujana, K. Y., & Yaswanth, D. (2022). Identification of medicinal plants using deep learning. *International Journal for Research in Applied Science and Engineering Technology*, *10*(4), 306–322. DOI: 10.22214/ijraset.2022.41190

Rashed, A., Feng, X., Prager, S. M., Porter, L. D., Knodel, J. J., Karasev, A., & Eigenbrode, S. D. (2018). Vector-borne viruses of pulse crops, with a particular emphasis on North American cropping system. *Annals of the Entomological Society of America*, *111*(4), 205–227.

Reddy, K., Patel, N., Singh, V. P., & Dasgupta, M. (2024). Root Knot Nematodes in Crops: Pathogen Dynamics and Mitigation Strategies. *Nematology Research Journal*, *12*(1), 45–60.

Ristaino, J. B., Anderson, P. K., Bebber, D. P., Brauman, K. A., Cunniffe, N. J., Fedoroff, N. V., Finegold, C., Garrett, K. A., Gilligan, C. A., Jones, C. M., Martin, M. D., MacDonald, G. K., Neenan, P., Records, A., Schmale, D. G., Tateosian, L., & Wei, Q. (2021). The persistent threat of emerging plant disease pandemics to global food security. *Proceedings of the National Academy of Sciences of the United States of America*, *118*(23), e2022239118. DOI: 10.1073/pnas.2022239118 PMID: 34021073

Rommens, C. M., Haring, M. A., Swords, K., Davies, H. V., & Belknap, W. R. (2007). The intragenic approach as a new extension to traditional plant breeding. *Trends in Plant Science*, *12*(9), 397–403. DOI: 10.1016/j.tplants.2007.08.001 PMID: 17692557

Rosso, P. H., Pushnik, J. C., Lay, M., & Stone, N. D. (2005). The role of archival weather data in Colletotrichum acutatum infection potential models for almond in California. *Plant Disease*, *89*(2), 131–139. DOI: 10.1094/PD-89-0131

Roy, A., Zhai, Y., Ortiz, J., Neff, M., Mandal, B., Mukherjee, S. K., & Pappu, H. R. (2019). Multiplexed editing of a begomovirus genome restricts escape mutant formation and disease development. *PLoS One*, *14*(10), e0223765. DOI: 10.1371/journal.pone.0223765 PMID: 31644604

Saakre, M., Jaiswal, S., Rathinam, M., Raman, K. V., Tilgam, J., Paul, K., Sreevathsa, R., & Pattanayak, D. (2024). Host-delivered RNA interference for durable pest resistance in plants: Advanced methods, challenges, and applications. *Molecular Biotechnology*, *66*(8), 1786–1805. DOI: 10.1007/s12033-023-00833-9 PMID: 37523020

Sachar, S., & Kumar, A. (2022). Deep ensemble learning for automatic medicinal leaf identification. *International Journal of Information Technology : an Official Journal of Bharati Vidyapeeth's Institute of Computer Applications and Management*, *14*(6), 3089–3097. DOI: 10.1007/s41870-022-01055-z PMID: 35975198

Sadhu, S., & Chatterjee, D. (2025). Advancing crop resilience: Plant breeding techniques to combat climate change challenges for small and marginal farmers. *Vigyan Varta*, *6*(3), 277–280.

Saikia, A. P., Hmangaihzuala, P. V. L., Datta, S., Gope, S., Deb, S., & Singh, K. R. (2021). Medicinal plant species classification using neural network classifier. In *2021 6th International Conference on Communication and Electronics Systems (ICCES)* (pp. 1805–1811). IEEE. DOI: 10.1109/ICCES51350.2021.9488971

Saleem, M. H., Potgieter, J., & Arif, K. M. (2019). Plant Disease Detection and Classification by Deep Learning. *Plants*, *8*(11), 2 ?"22. DOI: 10.3390/plants8110468 PMID: 31683734

Samal, I., Bhoi, T. K., Pradhan, A. K., & Mahanta, D. K. (2023). Plantix app: A success story of artificial intelligence in plant protection. *Van Sangyan, 24*.

Samal, I., Bhoi, T. K., Pradhan, A. K., & Mahanta, D. K. (2023). Plantix app: A success story of artificial intelligence in plant protection. *Van Sangyan*, *10*, 24–44.

Sánchez-Martín, J., Widrig, V., Herren, G., Wicker, T., Zbinden, H., Gronnier, J., Spörri, L., Praz, C. R., Heuberger, M., Kolodziej, M. C., Isaksson, J., Steuernagel, B., Karafiátová, M., Doležel, J., Zipfel, C., & Keller, B. (2021). Wheat Pm4 resistance to powdery mildew is controlled by alternative splice variants encoding chimeric proteins. *Nature Plants*, *9*(3), 502–503. DOI: 10.1038/s41477-023-01375-3 PMID: 36869130

Santillán Martínez, M. I., Bracuto, V., Koseoglou, E., Appiano, M., Jacobsen, E., Visser, R. G., Wolters, A.-M. A., & Bai, Y. (2020). CRISPR/Cas9-targeted mutagenesis of the tomato susceptibility gene PMR4 for resistance against powdery mildew. *BMC Plant Biology*, *20*(1), 284. DOI: 10.1186/s12870-020-02497-y PMID: 32560695

Saponara, S., & Elhanashi, A. (2022). Impact of image resizing on deep learning detectors for training time and model performance. In *Applications in Electronics Pervading Industry, Environment and Society: APPLEPIES 2021* (pp. 10–17). Springer International Publishing. DOI: 10.1007/978-3-030-95498-7_2

Savary, S., Ficke, A., Aubertot, J. N., & Hollier, C. (2012). Crop losses due to diseases and their implications for global food production losses and food security. *Food Security*, *4*(4), 519–537. DOI: 10.1007/s12571-012-0200-5

Schaart, J. G., van de Wiel, C. C. M., Lotz, L. A. P., & Smulders, M. J. M. (2016). Opportunities for Products of New Plant Breeding Techniques. *Trends in Plant Science*, *21*(5), 438–449. DOI: 10.1016/j.tplants.2015.11.006 PMID: 26654659

Scheben, A., Yuan, Y., & Edwards, D. (2016). Advances in genomics for adapting crops to climate change. *Current Plant Biology*, *6*, 2–10. DOI: 10.1016/j.cpb.2016.09.001

Schiemann, J., Dietz-Pfeilstetter, A., Hartung, F., Kohl, C., Romeis, J., & Sprink, T. (2019). Risk assessment and regulation of plants modified by modern biotechniques: Current status and future challenges. *Annual Review of Plant Biology*, *70*(1), 699–726. DOI: 10.1146/annurev-arplant-050718-100025 PMID: 30822113

Schuster, M., & Paliwal, K. K. (1997). Bidirectional recurrent neural networks. *IEEE Transactions on Signal Processing*, *45*(11), 2673–2681. DOI: 10.1109/78.650093

Seem, R. C. (1984). Disease incidence and severity relationships. *Annual Review of Phytopathology*, *22*(1), 133–150. DOI: 10.1146/annurev.py.22.090184.001025

Seipel, T., Ishaq, S. L., & Menalled, F. D. (2019). Agroecosystem resilience is modified by management system via plant–soil feedbacks. *Basic and Applied Ecology*, *39*, 1–9. Advance online publication. DOI: 10.1016/j.baae.2019.06.006

Seleiman, M. F., & Hafez, E. M. (2021). Optimizing inputs management for sustainable agricultural development. In *Mitigating environmental stresses for agricultural sustainability in Egypt* (pp. 487–507). Springer International Publishing. DOI: 10.1007/978-3-030-64323-2_18

Selvam, H., & Ragiman, S. (2024). *Eco-friendly Bt-based bioinsecticide for sustainable pest management*. Just Agriculture Newsletter.

Shafik, W. (2025c). Sustainable Agriculture and Diet in the Metaverse Era: Sustainable Farming and Nutritious Eating in the Metaverse. In *Food in the Metaverse and Web 3.0 Era: Intersecting Food, Technology, and Culture* (pp. 159–184). IGI Global Scientific Publishing.

Shafik, W., & Kalinaki, K. (2024). Societal and Ethical Implications of Technology-Enhanced Agriculture and Healthcare: An African Context. *2024 IST-Africa Conference (IST-Africa)*, 1–11. DOI: 10.23919/IST-Africa63983.2024.10569306

Shafik, W. (2024a). Barriers to Implementing Computational Intelligence-Based Agriculture System. In *Computational Intelligence in Internet of Agricultural Things* (pp. 193–219). Springer., DOI: 10.1007/978-3-031-67450-1_8

Shafik, W. (2024b). Biosensor-Based Drones Anomaly Detection Integration for Sustainable Agriculture Development. In *Achieving Food Security Through Sustainable Agriculture* (pp. 283–316). IGI Global., DOI: 10.4018/979-8-3693-4240-4.ch012

Shafik, W. (2025a). *Ecological Health Impact in the Marine World.*, DOI: 10.1007/978-3-031-82403-6_10

Shafik, W. (2025b). Sustainable Agriculture and Diet for Urban Development. In *Sustainable Smart Cities and the Future of Urban Development* (pp. 165–192). IGI Global Scientific Publishing.

Shafik, W., & Azrour, M. (2024). Building a Greener World: Harnessing the Power of IoT and Smart Devices for Sustainable Environment. In *Technical and Technological Solutions Towards a Sustainable Society and Circular Economy* (pp. 35–58). Springer., DOI: 10.1007/978-3-031-56292-1_3

Shafik, W., Tufail, A., Apong, R. A. A. H. M., & De Silva, L. C. (2024). Internet of Things for Smart Agricultural Practices. In *Internet of Things Applications and Technology* (pp. 190–217). Auerbach Publications., DOI: 10.1201/9781003458401-13

Shafik, W., Tufail, A., De Silva, C. L., Haji, R. A. A., & Apong, M. (2025). The Role, Application, and Impact of Artificial Intelligence in the Agriculture Industry. In *Future Tech Startups and Innovation in the Age of AI* (pp. 36–60). CRC Press.

Shafik, W., Tufail, A., Liyanage De Silva, C., & Awg Haji Mohd Apong, R. A. (2025). A novel hybrid inception-xception convolutional neural network for efficient plant disease classification and detection. *Scientific Reports*, *15*(1), 3936. DOI: 10.1038/s41598-024-82857-y PMID: 39890849

Shafik, W., Tufail, A., Liyanage, C. D. S., & Apong, R. A. A. H. M. (2023). Using a novel convolutional neural network for plant pests detection and disease classification. *Journal of the Science of Food and Agriculture*, *103*(12), 5849–5861. Advance online publication. DOI: 10.1002/jsfa.12700 PMID: 37177888

Shafik, W., Tufail, A., Namoun, A., De Silva, L. C., & Apong, R. A. A. H. M. (2023). A Systematic Literature Review on Plant Disease Detection: Motivations, Classification Techniques, Datasets, Challenges, and Future Trends. *IEEE Access : Practical Innovations, Open Solutions*, *11*, 59174–59203. DOI: 10.1109/ACCESS.2023.3284760

Sharma, A., Sharma, A., Kumar, R., Sharma, I., & Vats, A. K. (2021). PR proteins: key genes for engineering disease resistance in plants. In *Crop improvement* (pp. 81–98). CRC Press. DOI: 10.1201/9781003099079-7

Sharma, R.. (2024). Powdery mildew in plants: Epidemiology and control strategies. *Fungal Biology Reviews*, *58*(4), 312–328.

Sharma, S. (2024). Organic agriculture for sustainable food systems: A comprehensive review of benefits and constraints. *Turkish Journal of Agriculture - Food Science and Technology*, *12*(8), 1476–1481. DOI: 10.24925/turjaf.v12i8.1476-1481.6482

Shekhawat, U. K., Ganapathi, T. R., & Hadapad, A. B. (2012). Transgenic banana plants expressing small interfering RNAs targeted against viral replication initiation gene display high-level resistance to banana bunchy top virus infection. *The Journal of General Virology*, *93*(8), 1804–1813. DOI: 10.1099/vir.0.041871-0 PMID: 22552945

Shende, D., Thakare, L., Agrawal, R., & Wyawahare, N. (2023, August). Automatic disease detection for various plants leaf using image processing techniques and tensorflow algorithm. In International Conference on Electrical and Electronics Engineering (pp. 487-499). Singapore: Springer Nature Singapore.

Shende, D., Wyawahare, N., Thakare, L., & Agrawal, R. (2023, February). Design process for adaptive spraying of pesticides based on mutual plant health detection and monitoring: a review. In *2023 Third International Conference on Artificial Intelligence and Smart Energy (ICAIS)* (pp. 729-733). IEEE.

Shoaib, M., Shah, B., El-Sappagh, S., Ali, A., Ullah, A., Alenezi, F., Gechev, T., Hussain, T., & Ali, F. (2023). An Advanced Deep Learning Models-Based Plant Disease Detection: A Review of Recent Research. *Frontiers in Plant Science*, *14*, 1 ?"22. DOI: 10.3389/fpls.2023.1158933 PMID: 37025141

Shorten, C., & Khoshgoftaar, T. M. (2019). A survey on image data augmentation for deep learning. *Journal of Big Data*, *6*(1), 1–48. DOI: 10.1186/s40537-019-0197-0

Shukla, M., Al-Busaidi, K. T., Trivedi, M., & Tiwari, R. K. (2018). Status of research, regulations and challenges for genetically modified crops in India. *GM Crops and Food: Biotechnology in Agriculture and the Food Chain*, *9*(4), 173–188. DOI: 10.1080/21645698.2018.1529518 PMID: 30346874

Sindhu, C., &Bala Janani, B. (2023). Empowering plant disease resistance through genetic engineering: A thorough analysis of current approaches, challenges, and future prospects. *International Journal of Scientific Development and Research.*

Singh, B. K., Delgado-Baquerizo, M., Egidi, E., Guirado, E., Leach, J. E., Liu, H., & Trivedi, P. (2023). Climate change impacts on plant pathogens, food security and paths forward. *Nature Reviews. Microbiology*, *21*(10), 640–656. DOI: 10.1038/s41579-023-00900-7 PMID: 37131070

Singh, R., Shafik, W., Crowther, D., & Kumar, V. (Eds.). (2024). *Transforming Healthcare Sector Through Artificial Intelligence and Environmental Sustainability* (1st ed., Vol. 1). Springer Nature Singapore., DOI: 10.1007/978-981-97-9555-0

Singh, S., Sahoo, S., Mohapatra, R., Panigrahi, C. K., Daniel, G. R., Arunkumar, S., Sahoo, L., & Mohapatra, D. (2025). Advances in Genetic Engineering for Disease Resistant Crops: A Review. *Journal of Advances in Biology & Biotechnology*, *28*(6), 856–869. DOI: 10.9734/jabb/2025/v28i62447

Singh, V. K., Singh, R., Kumar, A., & Bhadouria, R. (2021). Current status of plant diseases and food security. In *Food security and plant disease management* (pp. 19–35). Woodhead Publishing. DOI: 10.1016/B978-0-12-821843-3.00019-2

Singh, V., & Misra, A. K. (2017). Detection of plant leaf diseases using image segmentation and soft computing techniques. *Information Processing in Agriculture*, *4*(1), 41–49. DOI: 10.1016/j.inpa.2016.10.005

Sisterson, M. S., & Stenger, D. C. (2013). Roguing with replacement in perennial crops: Conditions for successful disease management. *Phytopathology*, *103*(2), 117–128. DOI: 10.1094/PHYTO-05-12-0101-R PMID: 23075167

Sivaranjani, C., Kalinathan, L., Amutha, R., Kathavarayan, R. S., & Kumar, K. J. J. (2019). Real-time identification of medicinal plants using machine learning techniques. In *2019 International Conference on Computational Intelligence in Data Science (ICCIDS)* (pp. 1–4). IEEE. DOI: 10.1109/ICCIDS.2019.8862126

Smith, L. P. (1956). Potato blight forecasting by 90% humidity criteria. *Plant Pathology*, *5*(3), 83–87. DOI: 10.1111/j.1365-3059.1956.tb00093.x

Snape, J. W. (2004, September). Challenges of integrating conventional breeding and biotechnology: a personal view. In *Proceedings of the 4th international crop science congress* (Vol. 26).

Subbarayudu, C., & Kubendiran, M. (2024). A comprehensive survey on machine learning and deep learning techniques for crop disease prediction in smart agriculture. *Nature Environment & Pollution Technology*, *23*(2), 619–632. DOI: 10.46488/NEPT.2024.v23i02.003

Suhag, D. (2024). *Regulatory and ethical considerations in biotechnology. Handbook of Biomaterials for Medical Applications* (Vol. 2). Springer.

Sujatha, R., Krishnan, S., Chatterjee, J. M., & Gandomi, A. H. (2025). Advancing plant leaf disease detection integrating machine learning and deep learning. *Scientific Reports*, *15*(1), 11552. Advance online publication. DOI: 10.1038/s41598-024-72197-2 PMID: 40185756

Sukprasert, K. R. I. T. C. H. A. I., & Phadungkit, W. I. C. H. A. I. (2024). Integrating sustainable development frameworks into agricultural policies: A policy analysis perspective. *Contemp. Issues Behav. Soc. Sci*, *11*(2), 1–12.

Sun, K., Wolters, A.-M. A., Vossen, J. H., Rouwet, M. E., Loonen, A. E. H. M., Jacobsen, E., Visser, R. G. F., & Bai, Y. (2016). Silencing of six susceptibility genes results in potato late blight resistance. [Springer.]. *Transgenic Research*, *25*(5), 731–742. DOI: 10.1007/s11248-016-9964-2 PMID: 27233778

Sun, Y., Liu, Y., Wang, G., & Zhang, H. (2017). Deep Learning for Plant Identification in Natural Environment. *Computational Intelligence and Neuroscience*, *2017*, 1 ?"6. DOI: 10.1155/2017/7361042 PMID: 28611840

Suresha, M., Shreekanth, K. N., & Thirumalesh, B. V. (2017). Recognition of diseases in paddy leaves using KNN classifier. In *2017 2nd International Conference for Convergence in Technology (I2CT)* (pp. 663–666). IEEE. DOI: 10.1109/I2CT.2017.8226213

TA1/4rko Ylu, M., & Hanbay, D. (2019). Plant Disease and Pest Detection Using Deep Learning-Based Features. *Turkish Journal of Electrical Engineering and Computer Sciences*, *27*(3), 1636 ?"1651. DOI: 10.3906/elk-1809-181

Talakayala, A., Katta, S., & Garladinne, M. (2020). Genetic engineering of crops for insect resistance: An overview. *Journal of Biosciences*, *45*(1), 114. DOI: 10.1007/s12038-020-00081-y PMID: 33051408

Talasila, S., Rawal, K., Sethi, G., & Mss, S. (2022). Black gram Plant Leaf Disease (BPLD) dataset for recognition and classification of diseases using computer-vision algorithms. *Data in Brief, 45*, 108725.

Tandekar, D., & Dongre, S. (2023, July). Identification of various diseases in plant leaves using image processing and CNN approach. In 2023 14th International conference on computing communication and networking technologies (ICCCNT) (pp. 1-6). IEEE.

Tao, Y., Zhao, X., Mace, E., Henry, R., & Jordan, D. (2019). Exploring and exploiting pan-genomics for crop improvement. *Molecular Plant, 12*(2), 156–169. DOI: 10.1016/j.molp.2018.12.016 PMID: 30594655

Thakre, R. N., Kunte, P. A., Chavhan, N., Dhule, C., & Agrawal, R. (2023, November). UAV based system for detection in integrated insect management for agriculture using deep learning. In 2023 2nd International Conference on Futuristic Technologies (INCOFT) (pp. 1-6). IEEE.

Thomas, R. (2022). Legal Challenges in Biosafety of Transgenic Plants. *Issue 6 Indian JL & Legal Rsch., 4*, 1.

Thomas, S., Kuska, M. T., Bohnenkamp, D., Brugger, A., Alisaac, E., Wahabzada, M., Behmann, J., & Mahlein, A. K. (2018). Benefits of hyperspectral imaging for plant disease detection and plant protection: A technical perspective. *Journal of Plant Diseases and Protection, 125*(1), 5–20. DOI: 10.1007/s41348-017-0124-6

Thomson, J. (2021). *GM crops and the global divide*. Csiro Publishing. DOI: 10.1071/9781486312665

Tiwari, P., Bose, S. K., & Park, K. Il, Dufossé, L., & Fouillaud, M. (2024). Plant-Microbe Interactions under the Extreme Habitats and Their Potential Applications. In *Microorganisms* (Vol. 12, Issue 3). DOI: 10.3390/microorganisms12030448

Tonmoy, M. R., Hossain, M. M., Dey, N., & Mridha, M. F. (2025). MobilePlantViT: A mobile-friendly hybrid ViT for generalized plant disease image classification. *arXiv preprint*. https://doi.org//arXiv.2503.16628DOI: 10.48550

Too, E. C., Yujian, L., Njuki, S., & Yingchun, L. (2019). A comparative study of fine-tuning deep learning models for plant disease identification. *Computers and Electronics in Agriculture, 161*, 272–279. DOI: 10.1016/j.compag.2018.03.032

Trindade Maia, R., & Campos, M. A. (2021). Genetic variation: The source of biological diversity. *IntechOpen, 19*, 1–15.

Tripathi, J. N., Ntui, V. O., Ron, M., Muiruri, S. K., Britt, A., & Tripathi, L. (2019). CRISPR/Cas9 editing of endogenous banana streak virus in the B genome of Musa spp. overcomes a major challenge in banana breeding. *Communications Biology, 2*(1), 46. DOI: 10.1038/s42003-019-0288-7 PMID: 30729184

Tripathi, J. N., Ntui, V. O., Shah, T., & Tripathi, L. (2021). CRISPR/Cas9-mediated editing of DMR6 orthologue in banana (Musa spp.) confers enhanced resistance to bacterial disease. *Plant Biotechnology Journal*, *19*(7), 1291–1293. DOI: 10.1111/pbi.13614 PMID: 33934462

Trippa, D., Scalenghe, R., Basso, M. F., Panno, S., Davino, S., Morone, C., Giovino, A., Oufensou, S., Luchi, N., Yousefi, S., & Martinelli, F. (2024). Next-generation methods for early disease detection in crops. *Pest Management Science*, *80*(2), 245–261. DOI: 10.1002/ps.7733 PMID: 37599270

Trotsenko, O., & Slukin, S. (2020). Public-Private Partnerships for agricultural innovation: opportunities and challenges in world practice. In *E3S Web of Conferences* (Vol. 176, p. 06004). EDP Sciences. DOI: 10.1051/e3sconf/202017606004

Uddin, M. M., Islam, N., & Halimuzzaman, M. (2024). Genetic engineering in crop improvement for diseases on resistance. *Int J Sci Healthcare Res*, *9*(3), 262–276. DOI: 10.52403/ijshr.20240332

Uma, K. V., Bama, B. S., Sabarinathan, D., & Roomi, S. M. M. (2022). Identification and retrieval of medicinal plants of Southern India using EfficientB4Net. *Journal of Intelligent & Fuzzy Systems*, *42*(6), 5097–5112. DOI: 10.3233/JIFS-211426

United Nations Environment Programme. (n.d.). *How to feed 10 billion people*. Retrieved September 2, 2025, from https://www.unep.org/news-and-stories/story/how-feed-10-billion-people

United Nations. (2025). *The sustainable development goals report 2025*. Retrieved September 1, 2025, from https://unstats.un.org/sdgs/report/2025/

Valdez, D. B., Aliac, C. J. G., & Feliscuzo, L. S. (2022). Medicinal plant classification using convolutional neural network and transfer learning. In *2022 IEEE International Conference on Artificial Intelligence in Engineering and Technology (IICAIET)* (pp. 1–6). IEEE. DOI: 10.1109/IICAIET55139.2022.9936868

van Bruggen, A. H., Goss, E. M., Havelaar, A., van Diepeningen, A. D., Finckh, M. R., & Morris, J. G.Jr. (2019). One Health-Cycling of diverse microbial communities as a connecting force for soil, plant, animal, human and ecosystem health. *The Science of the Total Environment*, *664*, 927–937. DOI: 10.1016/j.scitotenv.2019.02.091 PMID: 30769316

Van der Plank, J. E. (1963). *Plant diseases: epidemics and control*. Academic Press.

Van Esse, H. P., Reuber, T. L., & van der Does, D. (2020). Genetic modification to improve disease resistance in crops. *The New Phytologist*, *225*(1), 70–86. DOI: 10.1111/nph.15967 PMID: 31135961

Varshney, R. K., Bohra, A., Yu, J., Graner, A., Zhang, Q., & Sorrells, M. E. (2021). Designing future crops: Genomics-assisted breeding comes of age. *Trends in Plant Science*, *26*(6), 631–649. DOI: 10.1016/j.tplants.2021.03.010 PMID: 33893045

Vaswani, A., Shazeer, N., Parmar, N., Uszkoreit, J., Jones, L., Gomez, A. N., & Polosukhin, I. (2017). Attention is all you need. *Advances in Neural Information Processing Systems*, *30*, 5998–6008.

Vincelli, P. (2016). Genetic engineering and sustainable crop disease management: Opportunities for case-by-case decision-making. *Sustainability (Basel)*, *8*(5), 495. DOI: 10.3390/su8050495

Vishnuoum. (n.d.). *Medicinal plant dataset (augmented)* [Data set]. Kaggle. https://www.kaggle.com/datasets/vishnuoum/medicinal-plantdataset-augmented

Wang, D., Mandal, P., Rahman, M. S., & Yang, L. (2025). Engineering tomato disease resistance by manipulating susceptibility genes. *Frontiers in Genome Editing*, *7*, 1537148. DOI: 10.3389/fgeed.2025.1537148 PMID: 39995605

Wang, Q., Qi, F., Sun, M., Qu, J., & Xue, J. (2019). Identification of tomato disease types and detection of infected areas based on deep convolutional neural networks and object detection techniques. *Computational Intelligence and Neuroscience*, *2019*, 1 ?"15. DOI: 10.1155/2019/9142753 PMID: 31933623

Wang, S., Xu, D., Liang, H., Bai, Y., Li, X., Zhou, J., Su, C., & Wei, W. (2025). Advances in deep learning applications for plant disease and pest detection: A review. *Remote Sensing (Basel)*, *17*(4), 698. DOI: 10.3390/rs17040698

Wang, Y., Yin, Y., Li, Y., Qu, T., Guo, Z., Peng, M., Jia, S., Wang, Q., Zhang, W., & Li, F. (2024). Classification of plant leaf disease recognition based on self-supervised learning. *Agronomy (Basel)*, *14*(3), 2 ?"13. DOI: 10.3390/agronomy14030500

Wani, K. I., Naeem, M., & Aftab, T. (2022). Chromium in plant-soil nexus: Speciation, uptake, transport and sustainable remediation techniques. In *Environmental Pollution* (Vol. 315). DOI: 10.1016/j.envpol.2022.120350

West, J. S., Bravo, C., Oberti, R., Lemaire, D., Moshou, D., & McCartney, H. A. (2019). The potential of optical canopy measurement for targeted control of field crop diseases. *Annual Review of Phytopathology*, *57*, 267–288. PMID: 12730386

Wezel, A., Herren, B. G., Kerr, R. B., Barrios, E., Gonçalves, A. L. R., & Sinclair, F. (2020). Agroecological principles and elements and their implications for transitioning to sustainable food systems. A review: Agroecological principles and elements and their implications for transitioning to sustainable food systems. A review. *Agronomy for Sustainable Development*, *40*(6), 40. DOI: 10.1007/s13593-020-00646-z

Yadav, S., Sarangi, S., Parmar, S., Thodusu, M., Soni, S., & Pandey, K. (2024). Genetic engineering in insect management: New frontiers in pest control. *Microbiology Research Journal International*, *34*(12), 106–121. DOI: 10.9734/mrji/2024/v34i121514

Yaduraju, N. T. (2021). The saga of genetically-modified (GM) and herbicide-tolerant (HT) crops in India. *Weeds-Journal of the Asian-Pacific Weed Science Society*, *3*(1), 38–55.

Yasin, E. T., Kursun, R., & Koklu, M. (2023, August). Deep learning-based classification of black gram plant leaf diseases: A comparative study. In 11th International conference on advanced technologies (ICAT'23), Istanbul-Turkiye.

Yin, K., Han, T., Xie, K., Zhao, J., Song, J., & Liu, Y. (2019). Engineer complete resistance to Cotton Leaf Curl Multan virus by the CRISPR/Cas9 system in *Nicotiana benthamiana. Phytopathology Research, 1*(1), 9. DOI: 10.1186/s42483-019-0017-7

Yuen, J., & Hughes, G. (2002). Bayesian analysis of plant disease prediction. *Plant Pathology, 51*(4), 407–412. DOI: 10.1046/j.0032-0862.2002.00741.x

Zadoks, J. C., & Schein, R. D. (1979). *Epidemiology and plant disease management.* Oxford University Press.

Zafar, K., Khan, M. Z., Amin, I., Mukhtar, Z., Yasmin, S., Arif, M., Ejaz, K., & Mansoor, S. (2020). Precise CRISPR-Cas9 mediated genome editing in super basmati rice for resistance against bacterial blight by targeting the major susceptibility gene. *Frontiers in Plant Science, 11*, 575. DOI: 10.3389/fpls.2020.00575 PMID: 32595655

Zaidi, S. S. E. A., Vanderschuren, H., Qaim, M., Mahfouz, M. M., Kohli, A., Mansoor, S., & Tester, M. (2019). New plant breeding technologies for food security. *Science, 363*(6434), 1390–1391. DOI: 10.1126/science.aav6316 PMID: 30923209

Zeng, X., Luo, Y., Vu, N. T. Q., Shen, S., Xia, K., & Zhang, M. (2020). CRISPR/Cas9-mediated mutation of OsSWEET14 in rice cv. Zhonghua11 confers resistance to *Xanthomonas oryzae* pv. oryzae without yield penalty. *BMC Plant Biology, 20*(1), 313. DOI: 10.1186/s12870-020-02524-y PMID: 32620100

Zhang, C., Chen, T., Chen, W., & Sankaran, S. (2023). Non-invasive evaluation of Ascochyta blight disease severity in chickpea using field asymmetric ion mobility spectrometry and hyperspectral imaging techniques. *Crop Protection (Guildford, Surrey), 165*, 106163.

Zhang, D., Hussain, A., Manghwar, H., Xie, K., Xie, S., Zhao, S., Larkin, R. M., Qing, P., Jin, S., & Ding, F. (2020). Genome editing with the CRISPR-Cas system: An art, ethics and global regulatory perspective. *Plant Biotechnology Journal, 18*(8), 1651–1669. DOI: 10.1111/pbi.13383 PMID: 32271968

Zhang, K., Wu, Q., Liu, A., & Meng, X. (2018). Can deep learning identify tomato leaf disease? *Advances in Multimedia, 2018*, 1 ?"10. DOI: 10.1155/2018/6710865

Zhang, M., Liu, Q., Yang, X., Xu, J., Liu, G., Yao, X., Ren, R., Xu, J., & Lou, L. (2020). CRISPR/Cas9-mediated mutagenesis of Clpsk1 in watermelon to confer resistance to Fusarium oxysporum f. sp. niveum. *Plant Cell Reports, 39*(5), 589–595. DOI: 10.1007/s00299-020-02516-0 PMID: 32152696

Zhan, X., Zhang, F., Zhong, Z., Chen, R., Wang, Y., Chang, L., Bock, R., Nie, B., & Zhang, J. (2019). Generation of virus-resistant potato plants by RNA genome targeting. *Plant Biotechnology Journal, 17*(9), 1814–1822. DOI: 10.1111/pbi.13102 PMID: 30803101

Zhou, J., Peng, Z., Long, J., Sosso, D., Liu, B. O., Eom, J. S., Huang, S., Liu, S., Vera Cruz, C., Frommer, W. B., White, F. F., & Yang, B. (2015). Gene targeting by the TAL effector PthXo2 reveals cryptic resistance gene for bacterial blight of rice. *The Plant Journal*, *82*(4), 632–643. DOI: 10.1111/tpj.12838 PMID: 25824104

Zin, I. A. M. D., Ibrahim, Z., Isa, D., Aliman, S., Sabri, N., & Mangshor, N. N. A. (2020). Herbal plant recognition using deep convolutional neural network. *Bulletin of Electrical Engineering and Informatics*, *9*(5), 2198–2205. DOI: 10.11591/eei.v9i5.2250

About the Contributors

Sana Basharat is working as a postdoctoral researcher at college of tropical agriculture and forestry at hainan university, haikou, hainan, China. She got her doctoral degree from Department of botany, university of agriculture faislabad, Pakistan. Her research direction involved structural and functional Strategies in Cenchrus Species to Combat Environmental Extremities Imposed by Multiple Abiotic Stresses. She had published more than 15 research articles in renowned journals including Plants, Pakistan Journal botany, Acta Physiologiae Plantarum, so on.

***

Dhananjay Bhagat is a dedicated academician and researcher currently serving as an Assistant Professor in the Department of CSE (AI), Vishwakarma Institute of Technology, Pune. He is pursuing his PhD in Computer Science and Engineering at the prestigious Visvesvaraya National Institute of Technology (VNIT), Nagpur. With a solid educational foundation with a Master's degree in Computer Science and Engineering from Sant Gadge Baba Amravati University, Dhananjay has developed a deep expertise in Artificial Intelligence, Machine Learning, Natural Language Processing, and Computer Vision. Over the years, Dhananjay has contributed significantly to the field through his research and publications. His work has been featured in several Scopus-indexed journals and conferences, covering diverse topics ranging from AI-powered resume optimization to deep learning applications in healthcare. He has also authored books on critical subjects like Business Intelligence and Cyber-Trolling Detection, reflecting his commitment to exploring the intersection of technology and society.

Manasvi Chand is pursuing BSc Biotechnology from Department of Life Sciences, Sharda University, Greater Noida. She has previous exposure in the field of Artificial Intelligence and Machine Learning through projects.

Deepak Gupta is a distinguished academician with over 20 years of teaching experience and expertise in Computer Science and Engineering. He holds a Ph.D. in Computer Engineering and has an extensive research portfolio, including many international journal papers, multiple conference proceedings, and several books. His research interests span machine learning, artificial intelligence, and data mining. Currently, Dr. Gupta serves as an Associate Professor and Head of Department of CSE (Emerging Branches) at ITM Gwalior, MP, India. He has also contributed significantly to the field through patents and copyrights and actively engages as a reviewer, conference chair and editor.

Swati Hazra obtained her joint M.Tech.-Ph.D. degree in 2021 from the Agricultural and Food Engineering Department, Indian Institute of Technology Kharagpur. She worked on the project "Development of Climatic Adaptive Agro-Technologies for Sustainable Wheat and Corn Production in Eastern India" under "Sustainable Food Security through: Technological Interventions for Production, Processing and Logistics" funded by the Ministry of Human Resource Development, Govt. of India. Dr. Hazra completed her B.Sc. (Hons.) and M.Sc. in Botany from University of Calcutta.

Ritesh Kalbande completed his Ph.D. research work at the Indian Institute of Tropical Meteorology, Pune while being registered at the Mohanlal Sukhadia University, Udaipur. The award of his doctorate degree is awaited. He worked on understanding the variability of volatile organic compounds (VOCs), identifying their sources and applying a machine learning model to explore ozone - VOCs relationship during Ph.D. Mr. Kalbande has obtained his Master of Engineering from the University of Cincinnati, USA, with Environmental Engineering as major while his Bachelor of Technology was from Visvesvaraya National Institute of Technology, Nagpur with Chemical Engineering as major.

Raju Naik is an Assistant Professor at Sri Krishna Devaraya College of Horticultural Sciences, Anantapuramu, India. He earned his Ph.D. in Genetics and Plant Breeding from Vasantrao Naik Marathwada Krishi Vidyapeeth (VNMKV), Parbhani, India. His research interests focus on drought stress studies in sorghum and multi-omics approaches, including genomics, phenolics, transcriptomics, proteomics, metabolomics, and ionomics. Dr. Naik has published research articles in international peer-reviewed journals and has presented his work at national and international scientific conferences.

Wasswa Shafik (Member, IEEE) received a Bachelor of Science degree in information technology from Ndejje University, Uganda, a Master of Engineering degree in Information Technology (Computer and Communication Networks) from Yazd University, Iran, and a PhD in Computer Science with the School of Digital Science, Universiti Brunei Darussalam, Brunei Darussalam. He is also the Founder and Research Director of the Dig Connectivity Research Laboratory (DCRLab) after serving as a Research Associate at the Network Interconnectivity Research Laboratory at Yazd University. Prior to this, he worked as a Community Data Analyst at Population Services International (PSI-Uganda), a Community Data Officer at the Programme for Accessible Health Communication (PACE-Uganda), a Research Assistant at the Socio-Economic Data Centre (SEDC-Uganda), Uganda, an Assistant Data Officer at TechnoServe, Kampala, Uganda, and Asmaah Charity Organization as Cheif Executive Officer. He has hundreds of publications with renowned publishers. His research interests include Artificial Intelligence, Neural Networks, Computer Vision, Smart Agriculture, Digital Health, and Green Computing.

Someshwar Siddi obtained Ph.D. from the Rayalaseema University, 2019. Having more than 20 years of experience in teaching the subject for UG and PG students affiliated to JNTUH. Presently working as Associate Professor, St. Martin's Engineering College, Secunderabad, India. Received awards from various Organizations in the filed of Research & Education. Published several research publications, Book Chapters in National and international journals. Published n number of patents. Successfully completed 16 courses from International Universities. Besides being an effective teacher, contributing to the institution in progressively fine tuning the teaching-learning processes, initiated several co-curricular and research activities to provide a platform for the students to pool together their individual talents across various disciplines.

Index

A

Agrobacterium 7, 56
Agroecosystem Resilience 92
and Accuracy 187, 202, 204, 233
Artificial Intelligence 30, 75, 93, 110, 111, 123, 125, 138, 141, 142, 143, 147, 166, 167, 196, 215, 216, 240

B

Bottle Gourd Leaf 220, 227, 237

C

Climate-Disease Models 96
Cloud 128, 132, 134, 135, 136, 148, 154, 155, 156, 162, 163, 164, 169
CRISPR 1, 2, 5, 6, 7, 8, 10, 13, 14, 15, 17, 18, 25, 26, 27, 28, 30, 31, 32, 33, 35, 36, 39, 41, 42, 49, 50, 51, 52, 53, 55, 57, 59, 60, 61, 62, 63, 64, 66, 68
CRISPR-Cas 52
Crop Diseases 140, 142, 221, 223
Crop Management 123, 125, 201
Crop Monitoring 194

D

Data Augmentation 133, 135, 137, 147, 149, 166, 238
Deep Learning 95, 99, 120, 121, 128, 130, 131, 136, 138, 140, 141, 142, 143, 146, 147, 148, 151, 152, 155, 156, 159, 162, 163, 164, 165, 166, 201, 215, 217, 219, 221, 223, 224, 225, 228, 231, 238, 240, 241, 242, 243
Disease Detection 93, 120, 121, 123, 124, 126, 128, 130, 132, 133, 136, 137, 140, 141, 142, 201, 203, 205, 206, 215, 216, 219, 221, 223, 224, 226, 233, 240, 241, 242, 243
Disease Forecasting 95, 96, 97, 98, 99, 100, 101, 102, 103, 104, 105, 107, 108, 109, 110, 111, 113, 114, 117, 118
Disease Resistance 1, 2, 3, 5, 6, 7, 8, 11, 12, 13, 14, 15, 21, 25, 27, 29, 30, 32, 35, 36, 40, 41, 42, 44, 50, 52, 57, 59, 60, 61, 62, 63, 64, 66, 69, 71, 73
Disease Suppression 16, 76, 82
DL 95, 127, 128, 130, 145, 146, 147, 148, 151, 152, 154, 155, 156, 159, 161, 162, 164, 224, 225

E

Early Warning Systems 19, 109, 113, 137
EfficientNet 134, 151, 152, 153, 154, 155, 156, 161, 162, 163, 165, 219, 221, 223, 231, 232, 233, 238

F

Functional Diversity 71, 75

G

Gene stacking 35, 41, 44, 49, 56, 60, 64
Genetic diversity 1, 2, 3, 9, 40, 41, 42, 72, 74, 75, 76, 79
Genetic Engineering 1, 2, 4, 5, 6, 7, 8, 11, 12, 14, 15, 17, 18, 20, 21, 25, 26, 27, 31, 32, 35, 36, 38, 39, 41, 42, 48, 50, 54, 55, 57, 59, 60, 66, 67, 69
Genome Editing 1, 3, 5, 6, 7, 8, 14, 18, 24, 25, 26, 28, 29, 33, 35, 49, 50, 55, 56, 57, 58, 61, 64, 66, 69
GMOs 11, 13, 35, 36, 39, 58, 68

I

Image Classification 120, 205
Image Resizing 166
Immunity 6, 19, 27, 39, 42, 43, 44, 45, 46, 47, 48, 50, 51, 55, 60, 61, 64, 66, 71, 73
InceptionNet 219, 221, 223, 231, 232, 233, 238

M

Machine Learning 19, 62, 95, 96, 97, 99, 100, 102, 103, 104, 105, 106, 107, 110, 111, 115, 116, 117, 119, 120, 125, 138, 140, 141, 142, 145, 147, 166, 167, 193, 194, 202, 204, 205, 206, 213, 240, 242
Medicine 28, 29, 143, 144, 145, 153, 158, 163, 165, 215
Microbial Communities 32, 71, 76

N

Natural Predators 73, 77

P

Pathogens 2, 3, 5, 6, 8, 9, 13, 15, 18, 20, 30, 36, 37, 38, 39, 40, 41, 42, 43, 44, 45, 47, 48, 49, 51, 56, 57, 59, 60, 64, 71, 73, 74, 76, 77, 78, 79, 80, 81, 85, 91, 96, 119, 124, 125, 137, 202, 203, 204, 205, 207, 208, 210, 212
Plant 1, 2, 3, 5, 6, 7, 11, 13, 19, 20, 21, 23, 24, 25, 26, 27, 28, 29, 30, 31, 32, 33, 35, 36, 37, 38, 39, 41, 43, 44, 45, 46, 47, 48, 49, 52, 55, 56, 57, 58, 59, 60, 61, 64, 66, 67, 68, 69, 71, 72, 73, 74, 75, 76, 77, 78, 79, 80, 81, 82, 83, 84, 85, 88, 90, 91, 92, 93, 95, 96, 98, 100, 103, 105, 106, 111, 119, 120, 121, 123, 124, 125, 126, 127, 128, 130, 131, 132, 133, 135, 136, 137, 138, 140, 141, 142, 143, 144, 145, 146, 147, 148, 149, 151, 152, 153, 154, 155, 159, 163, 164, 165, 166, 167, 200, 201, 202, 203, 204, 205, 206, 207, 208, 209, 210, 211, 212, 215, 216, 217, 219, 220, 221, 222, 223, 224, 225, 226, 227, 228, 231, 233, 238, 239, 240, 241, 242, 243
Plant Disease Detection 93, 120, 121, 124, 128, 130, 133, 136, 137, 140, 141, 142, 201, 203, 206, 233, 240, 241, 242, 243
Plant Diseases 1, 2, 3, 5, 25, 31, 36, 41, 71, 73, 77, 78, 81, 82, 95, 121, 123, 124, 125, 126, 127, 140, 142, 146, 200, 201, 202, 203, 204, 205, 206, 212, 219, 220, 221, 222, 223, 225, 228, 239, 241, 242
Plant Epidemiology 121
Plant-Pathogen Interactions 71, 125
Polyculture 73
Precision Agriculture 19, 22, 27, 110, 114, 118, 120, 125, 127, 138, 170, 193, 197, 200
Predictive Modeling 17, 18, 95
PR Proteins 6, 30, 45, 46, 47
Pulse Crops 200, 201, 204, 206, 212, 213, 216
Pumpkin Leaf 220, 225, 228, 229, 230, 234, 238

R

RNAi 1, 5, 6, 35, 39, 41, 48, 49, 54, 64
RNA Interference 1, 5, 6, 29, 39, 48, 49, 54

S

Sustainable Farming 27, 92, 195

T

Transfer Learning 110, 116, 131, 134, 135, 141, 146, 147, 167, 219, 221, 223, 224, 225, 234, 238

W

Web-Based Diagnostic Tool 199, 200, 213

www.ingramcontent.com/pod-product-compliance
Lightning Source LLC
LaVergne TN
LVHW081522300926
849161LV00006B/1264
* 9 7 9 8 3 3 7 3 2 4 2 9 6 *